MODEL GOD

THE
MEANING OF LIFE
SECOND EDITION

J.G. Lenhart

WITH COMMENTARY BY

Joel Swokowski

Copyright © 2023 by John Lenhart

All rights reserved.

No part of this publication may be reproduced, distributed, or transmitted in any form or by any means, including photocopying, recording, or other electronic or mechanical methods, without the prior written permission of the publisher, except as permitted by U.S. copyright law.
For permission requests, contact info@modelinggod.com.

Although the publisher and the author have made every effort to ensure that the information in this book was correct at press time and while this publication is designed to provide accurate information in regard to the subject matter covered, the publisher and the author assume no responsibility for errors, inaccuracies, omissions, or any other inconsistencies herein and hereby disclaim any liability to any party for any loss, damage, or disruption caused by errors or omissions, whether such errors or omissions result from negligence, accident, or any other cause.

All Bible scriptures are taken from the King James Version

Second edition
Commentary by Joel Swokowski

Original cover art and illustrations by Joe Bailen
Second edition cover art by Rob Warning
Book design by Mark D'Antoni, eBook DesignWorks

ISBN: 979-8-9856838-0-6

www.modelinggod.com

Table of Contents

Introduction. 8
Preface 12
   Joel Swokowski's Commentary. 19
Book 1 — Modeling God and Salvation **22**
   Part One — The Address. 23
   Chapter 1 — The Invitation 24
      Intentional Perspective 25
      The Excuse 26
      Avoiding Thought .27
      Summary 28
      Joel Swokowski's Commentary. 29
   Chapter 2 — God the Creator. 33
      Modeletics™ Principle #1: Non-Contradiction 34
      Definition of God the Creator. 36
      Free Will 37
      Joel Swokowski's Commentary. 41
   Chapter 3 — Righteous 47
      Modeletics™ Principle #2: Growth 49
      Modeletics™ Principle #3: Contrastive Thinking 50
      Summary 51
      Joel Swokowski's Commentary. 53
   Chapter 4 — Just 59
      Justice Applications 61
      Summary 63
      Joel Swokowski's Commentary. 65
   Chapter 5 — Applications 68
      Modeletics™ Principle #4: Causality 69
      Examples 71
      The Hard Question .72
      Progress 74
      Applications 76
      Part I Summary 78
      Joel Swokowski's Commentary. 81

Part Two — The Door 85
Chapter 6 — Faith 86
   "Golden Key" Faith 87
   Biblical Faith 88
   Experience and Understanding 89
   Knowledge, Understanding, and Wisdom 91
   Summary 93
   Joel Swokowski's Commentary 94
Chapter 7 — Faith Examples 96
   Little Faith 97
   Great Faith 100
   Summary 101
   Joel Swokowski's Commentary 102
Chapter 8 — Grace 105
   "Golden Key" Implications 106
   New Testament Grace 108
   Summary 110
   Joel Swokowski's Commentary 112
Chapter 9 — Grace Examples 115
   Grace Examples 116
   Understanding and Grace 117
   Action Grace 118
   Failing Grace 119
   Works 120
   Wisdom 123
   Summary 123
   Joel Swokowski's Commentary 124
Chapter 10 — Uniqueness 128
   Qualitative 129
   Quantitative 131
   Expectations 133
   Summary 134
   Joel Swokowski's Commentary 136
Chapter 11 — Salvation Process 139
   How Did God Justly Obtain Infinite Value? 140
   How Does God Give Us the Value? 141
   What is Sin? 142
   What Benefit is Confession and Repentance Before We Die? . . 144

Why Does Confession and Repentance Solve Our Problems? . . 146
How Does the World Try to Remove Guilt? 148
What is the Party? 149
Joel Swokowski's Commentary. 150
Chapter 12 — Salvation Implications 156
Salvation Summary 158
Grace Only 160
Evangelism 162
Joel Swokowski's Commentary. 165
Chapter 13 — Salvation and Rewards 168
Reward Model 170
Two Models 173
Joel Swokowski's Commentary. 177
Chapter 14 — The Way 180
Sanctification. 182
Jesus Model. 185
Joel Swokowski's Commentary. 186
Chapter 15 — Discipleship 188
Sacrifice. 190
Fundamental Christianity 193
Joel Swokowski's Commentary. 197
Afterword 201
Joel Swokowski's Commentary. 201

Book 2 — Christian Living, Modeling Life 204
Part One — The Journey. 205
Chapter 1 — Profitability 206
Progress. 207
Profitability 208
Two Philosophies 209
Bible Examples 211
Joel Swokowski's Commentary. 215
Chapter 2 — Truth 221
Truth 223
Implications. 224
Art 226
Summary 228
Joel Swokowski's Commentary. 230

Chapter 3 — Weed Killer	233
Resisting Evil	233
Model Building	235
Disagreement	236
Debate	237
Win, Win	239
Joel Swokowski's Commentary	241
Chapter 4 — Life	244
Three-Dimensional Definition	245
Eternal Life	247
Summary	249
Joel Swokowski's Commentary	250
Part Two — Pray & Love (Exchanging Value)	253
Chapter 5 — Prayer	254
Prayer	255
Witchcraft	256
Prayer Structure	257
Requesting a Result	258
Summary	259
Joel Swokowski's Commentary	260
Chapter 6 — Prayer Implications	268
Example	269
Implications	270
Praying About Others	273
Summary	275
Joel Swokowski's Commentary	277
Chapter 7 — Morals vs. Ethics	281
Getting a Value From Others	282
Guilt	284
Ethics	286
Joel Swokowski's Commentary	288
Chapter 8 — Are, Do, Have	291
Are, Do, Have	292
Have, Do, Are	293
Applications	294
Summary	296
Joel Swokowski's Commentary	296

Chapter 9 — Moral Code 301
   Changing Your Code 302
   Joel Swokowski's Commentary. 305
Chapter 10 — Determining Your "ARE" 306
   The Party 307
   The Antenna 310
   Techniques to Determine Your Uniqueness 311
   Modeling People. 314
   Summary 318
   Joel Swokowski's Commentary. 319
Chapter 11 — Love 322
   Exercise 323
   Definition 324
   Four Pillars 326
   Joel Swokowski's Commentary. 328
Chapter 12 — Love Applications 333
   Acting in Love 335
   Summary 336
   Joel Swokowski's Commentary. 340
Chapter 13 — Marriage. 344
   Emotion 345
   The Concentric Circle Theory. 346
   Change in Love 348
   Physical Are vs. ARE 349
   Summary 350
   Joel Swokowski's Commentary. 351
Chapter 14 — Marriage Applications 355
   Communication and Intimacy 358
   Joel Swokowski's Commentary. 361
Chapter 15 — Living the Life Philosophy 366
   Grace and Forgiveness 366
   Getting Out of Survival Mode. 367
   Hurdles vs. Drivers 368
   Physical vs. Spiritual 369
   Application 371
   Summary 373
   Joel Swokowski's Commentary. 375
Endnotes 379

Introduction

I've lived in Northeastern Wisconsin since I was four years old, way back in 1985. Before that, my upbringing revolved around being a pastor's kid. I was even born in Bob Jones University Hospital in Greenville SC. It may be that the Bible and Theology were as big a part of my past as they have become my present and seemingly will be my future.

I was blessed to have met J.G. Lenhart (the author of this book) in 2006, the year before its publication. I was even part of the group of people honored with the privilege of reading Modeling God when it was still in manuscript form. It changed my life in more ways than even I know.

In 2009, two years after Modeling God's publication, I started my journey towards Seminary. After finally submitting my life to the Lord a few years before, I immediately felt a calling into some type of ministry focused on teaching and preaching. This book was at the foundation of that journey. By the start of my schooling at Liberty University in 2009, I had already read and taught Modeling God multiple times. This book spoke to me; it seemed to be speaking right to me; it was speaking my language. I now had a grasp on who God is, what Salvation is, and how I can intentionally live the life God intended me to live. This was all possible because of the doctrine that was presented in this book.

My undergraduate classes (in which I double majored) and my graduate classes all had a heavy workload. However, due to my already having an understanding of the bedrock of what was being taught at university, the assignments were all a breeze. Yes, there was a lot of work. Yet I never struggled with having the answers and information I needed to complete the assignments. Because I had the answers to the foundation of all of Christianity, no matter the topic, the result was I had the answers to every subject. If I was given the assignment to do a research paper on Justification by Faith, I could do it because I understood the causes of Salvation. If I was given the assignment to do a research paper on God's Providence, I could do it because I understood who God is at the foundational level.

I quickly saw the different levels to which a person can experience the written word, especially the Bible:

1. there's *what* you read,
2. there's the specific meaning (*why*) in the context of what you read, and
3. there's the deeper meaning (*how*/Principle) that is true regardless of the context.

That deeper meaning, those contextless principles, is doctrine. This is exactly what Modeling God presents: the doctrine of the Bible.

When this is embraced, it actually gives the reader the ability to understand the Bible in an entirely new way, in the way God intended. It starts to make sense. It starts to become applicable in our modern day and age. It starts to help a person do God's Word intentionally. It starts to help a person hear His still, small voice clearer than ever!

This also makes Modeling God a book that differs from what people are used to. This is what caused so many people to stand against J.G. Lenhart and his book. Modeling God went out of print due to attacks stating that the author was a heretic. However, as of the date of this publication, no one was ever able to explain to the author what was specifically heretical and give the correct explanation to be used in place of the "heretical" belief. Realize, everyone shied away from this opportunity and participated in whispering (gossip) because if their explanation was wrong, they would be the heretic. Even the most vocal opponents to this work had to factually misquote this work over 200 times in an attempt to prove their point. Basically, opponents used *ad hominem* and *ad populum*; approaches that have been declared as logical fallacies for millennia.

This is a dense textbook. Even when it's read from a humble heart, the way it's written and the content of this book can be hard to digest. Take your time. I'm here to help make this feel like less of a textbook. I'm here to help you digest this information so it can bless your walk with God the way it did mine and so many others.

If you consider yourself an "expert" with respect to Christianity and/or the Bible, it may take you twice as long to understand this book as you will first have to unlearn certain contradictory beliefs before you can learn the non-contradictory explanations from the Bible. If this is you, you will also feel what the critics of this book felt and I'd like to help you avoid committing the sin of whispering.

This second edition is in ebook form along with my commentary at the end of every chapter. The commentary may serve several purposes depending on the chapter, including:

1. Clarity and insight.
2. An answer to critics.
3. Provision of historical context and applications.
4. Additional information realized since the first publishing.

Modeling God was published fifteen years ago and has not only stood the test of time but has also explained what we have seen in the church since that time (pastors being burnt out, lower attendance, etc.). Each time I read this book, I still learn, it still seems fresh, and it's still needed. I hope you have an enriching journey. God bless you!

Joel Swokowski, 2022
Senior Pastor, Music of Life Church – Kimberly
Master's Degree in Theological Studies, Liberty University Baptist Theological Seminary
Bachelor's Degree in Religion, Liberty University
Bachelor's Degree in Education, Liberty University

Preface

The first of these two books identifies four tools for determining a comprehensive worldview that presents and proves the only possible explanation for a supreme being and salvation. The second book applies this worldview to everyday interactions and presents the meaning of life.

The term "worldview" has a long history, but is a relatively new term for most people and refers to an overarching framework or model that gives us a way of looking at and understanding the world. In a culture, it is expressed across the board in areas such as art, politics, science, and religion.

Actually, everyone has a worldview. Everyone has a personal model that explains how he or she views the world. In fact, part of the task of evangelism is to free people from the power of false worldviews by diagnosing points where they contradict reality. The result of worldview thinking is a deepening of our spiritual character and the character of our lives. Ultimately, the worldview brings consistency to a message and meaning to the existence of the individual.

Recently, numerous books have been written on the importance of getting the Christian community to present themselves in terms of a worldview. Currently, Christianity is presenting itself in piecemeal fashion; that is, they are treating each issue (e.g. abortion) as a separate challenge to their moral structure. Once the issues have been compartmentalized, it is relatively easy for opponents of Christianity to show the general public the contradictions between the issues. The result is that Christianity is represented as not having real-world, practical applications to the individual, as well as our culture.

Rather than allowing outside forces to divide and conquer the Christian influence on our country and world, Christians need to educate people as to how each issue is actually a part of a non-contradictory, superstructure belief. The reason this hasn't been done is that a non-contradictory model that non-Christians can understand hasn't been identified. This is the crucial challenge. Currently, Christians don't extend their explanation of their worldview past creation ("God created the heavens and the Earth"), the fall ("man is sinful"), and redemption ("Jesus died for our sins"). This traditional worldview has been good enough to differentiate Christianity from the other well-known worldviews because this simple model shows them the contradictions in these established religions.

Today, however, their model needs to apply to real-world topics. These real-world topics are subtle and opinions are derived from the worldviews of common, everyday people. Since they are addressing subtle worldviews, their model needs to be more detailed and objective. This will require a significantly more complex model that makes sense to non-Christians. This book begins the presentation of this long-awaited model.

The goal of this series is to present a model that explains how God operates, and how the concepts people talk about when expressing their beliefs fit into the worldview. The order of explanation, beginning with the first book, can best be explained by imagining a wall that is 200 miles long and 50 miles high. The wall is made up of tiles that are hundreds of feet square.

Most people are within arm's length of the wall. They know they are looking at a blue tile, for instance. However, they are telling each other that the wall is completely blue. In fact, there are people walking up and down the blue section explaining to the people more detailed facts about the blue wall.

If one were to leave the comfort of the crowd and back several hundred feet away from the wall, they would notice that they were looking only at a blue tile. They would realize the wall is not entirely blue but made up of different colors. Everyone at this distance would feel superior to the others because they realize the wall isn't entirely one color.

If one continued moving away from the wall, they would find the vast number of tiles overwhelming. It would appear the wall is nothing more than a random collection of colored tiles. At this point, the individual must make a hard decision—continue this uncomfortable journey away from the wall or return to the comfort of the crowd.

If they persevered and moved further from the wall, they would find the tiles make up a picture of a man standing on grass. It now appears the wall is a mural and is actually trying to portray an image. As they encounter people who have walked parallel to the wall, or read what others have learned, they find people talking about a fish out of water. In an effort to agree, they think they must be speaking figuratively and have interpreted the man to be the fish out of water; however, if they were to decide to get as far away from the wall as possible in order to take in the big picture, they would eventually find the mural is of a person teaching a man how to fish.

Most people focus on becoming experts on the details of a belief system. They walk closer to the wall. These books are attempting to take you far enough away from the wall that you get the big picture. This first book will only show you the wall isn't entirely blue. As you make your way through the books, the concepts in the first two books should begin to present an image. No doubt you will notice some tiles are missing, however, not enough to take away from the overall image. I will leave the filling in of these tiles for subsequent books.

In identifying this worldview, I found the key to every person's belief lies in the definition of a set of key words. It seemed to be impossible for anyone to describe their beliefs without using one of these key words. I realized the confusion we experience is due to the contradictory definitions of these key words.

These key words are the tiles in the wall. Consequently, the worldview depends on defining these key words in a non-contradictory fashion. In order to identify specific definitions, I applied SI Hayakawa's "Ladder of Abstraction" to these key words (e.g., God and salvation).

The classic explanation involves Bessie the Cow. The most specific word to describe her is "Bessie." The word "Bessie" can be seen as appearing on the lowest rung of the ladder. The next rung up could be "Cow." This word does refer to "Bessie," but it can refer to "Elsie" as well. Continuing up the ladder, the next rung could be labeled "Livestock." The subsequent rung could be labeled "Farm Assets." Finally, for our example, the last rung could be labeled "Wealth." All of these words apply to Bessie.

When we want to be inclusive, we move up the ladder. For instance, if I want to talk about Bessie and you want to talk about Elsie, it is easier for us to talk about Cows. The farmer could have a discussion with the CEO of a multinational corporation if they discussed Wealth. Moving up the ladder makes us more comfortable with others.

However, if we want to solve a problem, we need to become more specific. We need to move down the ladder. In fact, one could say we need to become more intentional. Failures to communicate occur when people operate on different rungs of the ladder.

If someone talked to me about Bessie and spoke of a Farm Asset, I may ask whether they were going to use unleaded or diesel fuel in his Farm

Asset. When they say they are talking about Livestock, they have moved down the ladder. However, I may then ask him what kind of saddle they have for his livestock. At this point, they would talk about Cows. We would be unable to have a non-contradictory discussion until they moved to the bottom rung.

I began to notice other examples of people speaking on different rungs. For instance, this is what occurred when Jesus spoke of "eating his flesh" (John 6:54) or "living waters" (John 4:10). We will see Jesus did this to determine who wanted to be intentional. On the other hand, some politicians do this to create confusion. Unfortunately, I find most clergy also do this when they give a sermon.

Actually, I found that all people (including myself) refer to the key words with definitions that are three or four rungs up the ladder. This is because we all have contradictory worldviews and need to cling to definitions at this level of abstraction to remain comfortable. In order to identify a non-contradictory model, I had to get all of these words defined at their bottom rung. Once I had these definitions, I discussed the worldview with theologians, clergy, professors, Christians, and atheists.

For the last five years, no one has been able to find a contradiction. In fact, the proof of the accuracy of the model has been the ability to show others how they can intentionally increase their faith or righteousness in the next five minutes. After all, if you understand the bottom-rung definitions of the key words (e.g., "faith" and "righteous"), you will know how to intentionally increase each attribute.

Finally, we will find that once we complete the worldview and have addressed the last concepts, we will end up where we began in the first book. The proof that this worldview is non-contradictory occurs when these last concepts mesh with the very first concepts. In effect, we have modeled theology.

Theology is the study of God. The study of the key concepts has a natural progression that has been recognized from the beginning. For example, every book on theology that discusses salvation must immediately address faith, grace, rewards, and free will.

I see the entire worldview as a series of ladders in a circle. Each ladder addresses a specific key word (e.g., grace). The rungs on each ladder correspond to a definition of each key word. Referring to the Ladder of Abstraction, the rungs at the bottom are the non-contradictory definitions. The rungs above increase in abstraction as you go further from the bottom rung.

In addition, lines connect this circle of ladders. Each concept is naturally connected to another concept. As I mentioned previously, any discussion of salvation leads to a discussion of grace, faith, free will, and rewards.

The goal of these books is to define all the key words at the bottom rung. When this is done, the contradictions will be removed. The proof is when we start at God and work our way to the Meaning of Life, we will end up back to God. At that point, our definition of the Meaning of Life should coincide with our definition of God and all the other key words.

If the definition of the Meaning of Life was at a higher rung, then we'd need to continue to work out the contradictions. The reason there could be a disconnect between the Meaning of Life and God is when you go from one concept to one it is connected to, you will only be able to match the rung level or go higher on the next word's Ladder of Abstraction.

When a person gives a definition of a key word, there are two ways to expose their contradictions. The first and easiest way is to ask questions until they change their definition (moves down the ladder). Since these books will help you know the definition at the lowest rung, it is pretty easy to ask the right questions to expose his contradictions.

The second way is to ask questions about the ladder that is connected to the key word being discussed. As mentioned previously, in order to avoid contradictions, the person will have to stay at the same level or become more abstract. They cannot define the connecting concept in a less abstract manner. If you keep moving to the next ladder, they will eventually drift up to a definition that is ridiculous.

The history of theology is filled with people who gave a definition on a higher rung and eventually ran out of steam as they applied their definition to the next key word. Defining the connecting concept less abstractly exposes the contradictions in the first concept. Clearly, the only non-contradictory theology would be the one made of definitions solely from the bottom rung of each ladder.

This first book focuses on the modern apologetic tools necessary to identify the worldview (Modeletics™) and the application of these tools to identify non-contradictory models for God and salvation. I call these "Foundational Principles," because these are the principles every person should understand before they begin their spiritual journey.

The first five chapters introduce Modeletics and use these tools to determine the model for God. The next five chapters introduce the principles of the Salvation Model and show that grace cannot be "unmerited favor." The final five chapters complete the Salvation Model and show that Jesus actually had two messages:

1. How to get to heaven (Salvation Model) and
2. How to get rewards on earth and in heaven (Reward Model).

It seems every other book about Jesus (except for the Bible) tries to resolve these two messages into one by ignoring the verses connected with the other message, abstracting Christ's words until they apply to both, or arguing for interpretation/translation errors.

Joel Swokowski's Commentary

The "four tools" mentioned in the first sentence are "God-given" tools. As we go, I'll give a scriptural reference for each of the tools, specifically showing how Jesus used these tools Himself. He didn't teach them as if they were new, He used them as if and because they've always existed.

The Preface discussed the need for Christians to have explanations for real-world topics. This book gives those explanations and this is why this book and the information within are still relevant and needed today.

As stated in my introduction, this book presents doctrine. These are the foundational truths/principles that are true regardless of what you're reading and regardless of the context. You can apply this to the "Ladder of Abstraction" as well in that the "bottom rung" of any ladder would be the Doctrinal explanation of that concept.

The wall analogy is a great example of analysis vs. synthesis. Analysis involves breaking something down to its smallest parts. Today, people realize this doesn't give us any answers related to purpose because you eventually end up with nothing. In Europe, they drive on the left side of the road. Will taking a European car apart explain why they drive on the left side of the road? No. The understanding and purpose always lie outside of the thing we are looking at, which is synthesis. The author rightly chose synthesis over analysis to understand the Bible and the worldview which is the first reason why he was able to come to this model and write this book when no one else has been able to.

The author used the King James Version of the Bible to show this model was determinable for hundreds of years. Furthermore, we will see at the end of Book 1, CS Lewis told us to determine a Model for God!

Key Terms

Worldview: a model that gives us a way of looking at and understanding the world. This gets even simpler when you treat it like a simple compound word: world & view. In other words, I can define this merely as, "how I view the world." What this book does is give some of the key words and their definitions that every person needs in order to intentionally understand their worldview. For example, whether a person believes in God or not, they do need to account for "God" in their belief about "how we all got here" or "how this all started."

Model: an explanation for observed effects. A model for any concept helps a person understand the causes of that concept and the resulting effects. We all have a model for every belief we carry, whether we can explain it or not. Can you believe that one of the ways the author was persecuted was for using the term "model?"

Models are used in everyday life. For example, an architectural blueprint of a building helps the contractors understand the foundational elements of the building that guides the placement of all the rest of the components of that building. The blueprint is a model. That reminds me! Didn't God give the Israelites a "blueprint" to create the Tabernacle in Exodus 25:9? He called it a "pattern" and that word means "model!" Even God uses models!

You can see the models presented in this book in the same way, a blueprint. The models will give you the foundational elements (doctrine/*how*) that will guide you in the placement of all the components (*what* & *why*) of the Bible.

Furthermore, when you look at the Greek word for "model," the closest English translation to model is "relationship." This brings even more clarity to what it means to have a "model" for any belief or concept you hold. The model would attempt to show the relationship between what we see in the tangible realm and the intangible elements that are the source of what we observe.

From what I presented in the Introduction, a model would be representative of the "relationship" between:

1. *What* you read,
2. The specific meaning (*why*) in the context of what you read, and
3. The deeper meaning (*how*/Principle) that is true regardless of the context.

Principle: A foundational truth upon which others are built. You'll see this book is chock-full of these. Be prepared to be introduced to a plethora of words. But don't worry, they'll all be clearly defined and explained.

Unmerited Favor: the traditional and contradictory definition for the doctrine of Grace. As you read, you'll see how this definition cannot be true. This happened to be one of the points in this book that was the most hotly contested. Yet, no one could ever give the right answer to the definition of Grace without relying on man-made tradition.

Book 1

Modeling God and Salvation

PART ONE

The Address

INSIDE

Chapter 1: The Invitation 24

Chapter 2: God the Creator 33

Chapter 3: Righteous 47

Chapter 4: Just 59

Chapter 5: Applications 68

CHAPTER 1

The Invitation

IMAGINE YOU ARE invited to a party where you will be able to participate in any sensual pleasure you desire for as long as you want.

It's an exclusive party, but the invitation I hand you simply says "Admit One" and has a place for you to fill in your name. As you look at the invitation, you notice it contains no other information. The first question you would probably ask is, "When is the party?"

You are relieved to find out the party was several months away and you had plenty of time to fit it into your schedule. Your next question would probably be, "Where is the party?"

To this, I might say, "Chicago."

As I turn to leave, you stop me so you can ask, "Where in Chicago?"

When I answer, "In a building with a sign in front. Why do you ask?" you may begin to show frustration.

"Why do I ask? I want to know how to find it, that's why!"

My answer would depend on where you are coming from. If you are a great distance away, I'd tell you to first take a plane. Then, I'd tell you to drive a car. As you try to interrupt, I say, "Let me be more helpful—when you drive the car, one pedal is the gas. That makes the car go. The other pedal is the brake. That makes the car stop. There is also a steering wheel. That helps you direct the car."

If you haven't given up, you may ask, "What am I supposed to do, drive every street until I find it? Could you please give me an address?" The reason you want an address is to know exactly where the party is so that you can find it intentionally. Another benefit of an address is that it allows you to measure your progress.

Up until you get the address, this process would be totally unacceptable for directions to the party. However, these are the same type of directions we get from people when we ask how to get closer to God. Think about it—the most important desire of your heart and you settle for directions that you'd consider ridiculous for something less important.

Intentional Perspective

When we ask how to get closer to God, we are told to read the Bible, pray, believe (have faith), and love others. What parts of the Bible should we read? Read the whole thing? How is this any different from me giving you a Chicago phone book and saying, "The address is in here"? In the same way that we are looking for a specific street; when we read the Bible, what are we supposed to focus on?

How do we build our faith or love more? People usually point to Hebrews 11:1 ("Now faith is the substance of things hoped for") or 1 Corinthians 13:4 ("Love is long-suffering, kind," etc.) to answer these questions, yet these passages don't tell us how faith or love works or, more importantly, how to intentionally increase these attributes.

Unfortunately, most people believe that when it comes to religion, we can't be intentional. They express this by saying, "We can't know these things," or, "We aren't supposed to try to understand them; it's a mystery." Some would even say it is "unnatural" to be able to explain these principles. Let's deal with the first two responses separately.

First, religion commands us to make progress in these areas. In fact, we are held accountable for it. If we "can't know" how to intentionally make progress, then why are we told to do these things? More to the point, why are we guilty when we don't grow in these areas?

The weaker of the two responses is that we aren't supposed to try to understand. This goes against everything we do, from reading the Bible to listening to a sermon. What is the purpose of teachers if it isn't to try to intentionally help others understand? What is the purpose of the Bible if God doesn't want us to try to understand? More importantly, why do these people want to avoid thinking?

The Excuse

These questions make some people so uncomfortable, sometimes their response is, "You are being scientific, while I have an artistic perspective." When it is posed this way, this is an excuse. This has nothing to do with being creative.

When they say a person is scientific, they are saying that person believes if they understand the cause, they can intentionally get a desired result. When they say they are artistic, they are saying they don't do things intentionally because they aren't supposed to or can't know the causes. These are the two previously mentioned responses. Their real hope is to have an excuse for why they don't have to think. They believe this will allow them to do whatever they want without guilt.

Isn't it interesting that unintentional people get upset when others become unintentional? They get mad when their car breaks down, their medication doesn't work, or their leaders make bad decisions. Yet, intentional people don't get upset when others become intentional. In fact, they can be intentional in everything they do, while people with the excuse cannot be unintentional in everything they do.

If you talk to these artists long enough, you will find they are very intentional with their art. They have certain guidelines on what does and doesn't work. You will see this very quickly if you disagree with them. After all, if things can't be understood, how can they be so sure their point of view is correct? In fact, their ability to convey the initial point of being unintentional requires them to be intentional in order to be understood. For example, they must be intentional when they use words. In reality, it takes an intentional approach to remain unintentional.

Later, we will see that the areas they want to remain unintentional in are areas in which they are comfortable. Their goal is to have an excuse not to think.

Avoiding Thought

People who avoid thinking believe that not understanding somehow relieves them from the responsibility of changing their behavior. As you'll see, however, the concepts in this book are true and impact our lives whether we understand them or not. What if we followed their example in other areas?

For example, over a hundred years ago, Dr. Ignaz Semmelweis noticed that maternity patients were dying at an alarming rate. He made the connection that students who had just finished working on cadavers were treating the patients! After insisting that the students wash their hands before treating the mothers, the number of deaths decreased by over a factor of five.[1]

Should Dr. Semmelweis have not tried to understand the reason maternity patients were dying? Should he have not altered the behavior of his students? Most importantly, even if he had decided he couldn't understand, it still would have negatively affected his patients. Allow me one more analogy…

Wine is made from fermented grapes. People initially found that when grapes were crushed and allowed to stand, the juice unexpectedly turned into alcohol. Eventually, they understood that yeast on the skin of the grape converted the sugar in the grape juice into alcohol. What was their response? Was it to continue crushing grapes and hope for the best? No; they began to apply this understanding and intentionally add yeast to the crushed grapes in order to ensure progress toward turning the juice into wine.

Summary

God acts according to a set of principles whether we understand them or not. If we don't understand them, things appear to be random and we will eventually become frustrated. However, if our understanding is increased and applied, it will result in more progress toward improving our interactions with God. This is most helpful when we face adversity. It is at these moments when everyone strives to be intentional. God knows that and wants to help us overcome those times intentionally. He has given us everything we need in order to intentionally grow and overcome adversity, but we must understand how the principles work if we want to ensure our progress, especially through the tough times. Ultimately, understanding God from first principles involves thinking more, not less.

We can know how God works. In fact, we are supposed to know. God did give us an address because He wants us to make it to His party.

- Do you believe you can know how to intentionally get closer to God?
- Do you believe you are supposed to know how to intentionally get closer to God?
- How does it make you feel when people give you abstract instructions like, "Read your Bible"?
- What do you believe is the biggest barrier preventing you from getting closer to God?
- What information do you think you need in order to get closer to God?

Joel Swokowski's *Commentary*

This chapter sets the precedence of being intentional, which is the heart of this book. This book is meant to help you intentionally understand God, salvation, the meaning of life, etc. This book is meant to help you understand these concepts at the level where you can live them out on purpose or, in other words, intentionally!

God has given us a guidebook: His Word, the Bible. So then, why do so many Christians fear whether or not they're really saved? Why do so many Christians say and believe that God is a mystery, that He can't be understood? There are so many questions that Christians have that God has given answers to, yet the church seems blind to these answers.

It seems to me that God wants to be known. What do you believe? Does God want you to know Him? Or is God preventing you from knowing Him?

> "Thus saith the Lord, Let not the wise man glory in his wisdom, neither let the mighty man glory in his might, let not the rich man glory in his riches: But let him that glorieth glory in this, that he understandeth and knoweth me, that I am the Lord which exercise lovingkindness, judgment, and righteousness, in the earth: for in these things I delight, saith the Lord" (Jeremiah 9:23-24).

The Bible shows us very clearly that God *wants* to be known!

God Wants to be Known

The verses from Jeremiah 9 have great meaning in my life. Early on in my walk with the Lord, I was given the opportunity to attend a Leadership Workshop with other Christian leaders. I was on fire for the Lord, with others who appeared to feel the same. These were people in full-time ministry seeking to teach and learn more about our Father in Heaven. It was a treat for an aspiring teacher.

I remember vividly the last day of class. I had already read Modeling God and had been applying and sharing the truth within this book. I was eager and zealous to give others this truth as well. I wanted other people to be transformed in the way I was. I shared with a leader what I knew to be God's Nature. I defined it for her and started to expose the contradictions that I had been hearing in the church. Her response baffled me. She accused me of pride and spiritual abuse! She claimed that I could not know God the way I was saying I did and that her own grandfather, on his deathbed, did not know the Lord the way I was saying was possible.

I now know that she was the one spiritually abusing me. She negated my beliefs and instead of bringing clarity to what I was sharing, she brought confusion and frustration.

I wonder, how many other Christians have had the same experiences I had? What is your definition for God's Nature?

The verses from Jeremiah 9 come into play in this story by way of my brother. At the time, my older brother was mentoring me and had actually been part of this Leadership Workshop that I was attending. He saw the entire exchange and he saw how distraught I was at the hands of a leader who was supposed to be nurturing and caring towards me. That same afternoon, my brother shared Jeremiah 9:23-24 with me. He encouraged me to continue my pursuit of understanding God. He had my back. I'm not sure what would have happened to me if I did not have the support I did from my brother who helped me jump over the stumblingblock that was put before my path that day.

Discussion Questions

You probably also noticed that this chapter introduced some discussion questions. I've revisited these questions in my own life every time I've read this book. I've also used them as a guide in discussing these topics and questions with others in the Modeling God classes that I've been privileged to lead.

My advice to you is to answer each question, either by writing them out or speaking them out loud. We're all on different paths and we're all making progress at different speeds. Answering these questions can help you mark your own progress in your journey of reading this book. Furthermore, it's a benefit to your own thought process to answer these questions for yourself. You may find that some of the harder, more convicting things you learn may cause your mind to start reeling. It may even cause you to have a hard time falling asleep. When I experience this happening in my life, I take it as a sign that I need to state my own beliefs about what I'm learning.

Jesus tells us, "But let your communication be, Yea, yea; Nay, nay…" (Matthew 5:37) and also to go hot or cold, and not be lukewarm (Revelation 3:16-17). Answering the questions at the end of each chapter can be one way for you to clearly and boldly state what you believe, to go hot or cold, and just maybe see what God will do on your behalf. Let's be active and intentional readers!

CHAPTER 2

God the Creator

IF WE ARE trying to measure progress, then we must define the goal we are moving towards. In order to tell if we are making progress in our relationship with God, it would make sense to first understand who or what God is; we need to know God's address.

To begin with, we know God the Creator is not a human (physical being). God is a spirit. What this means is that while God is very real, God is not tangible in a physical sense. The only thing we know of that is both "real" and "not tangible" are ideas, or more to the point, principles.

God is a set of non-contradictory principles. Some might argue that Jesus is God, and He was definitely tangible; however, Jesus was the physical embodiment of God—He was the physical embodiment of these non-contradictory principles. Before we discuss the non-contradictory principles, let's look at contradictions.

Modeletics™ Principle #1: Non-Contradiction

Contradictions don't exist; conflicts do. For instance, I can be both happy and sad. This is a conflict. However, I cannot be both happy and not happy at all. That would be a contradiction. People intuitively believe that contradictions don't exist. They prove it every time they argue.

Every time a person claims to be right and proves you're wrong, what they are actually doing is showing where your statements contradict themselves. Therefore, the fact that contradictions don't exist is a given, it's accepted without mentioning. Next time someone tells you they believe contradictions exist, wait for them to show you are wrong and just say, "So? I can't be wrong because contradictions don't exist."

Non-contradiction is the first foundational principle that has been proven since the beginning of time. In Part I, we are going to cover the four principles that make up the foundation for every reasoning, interpretation, and logic tool that exists today. For instance, apologetics are based on these foundational principles. I have called these principles "Modeletics™" because we are applying these modeling tools in order to determine the non-contradictory worldview. These are the same tools, however, used to solve problems and increase understanding in every other field (science, psychology, economics, etc.). The limitations of every other reasoning tool can be corrected by these four foundational principles.

Now, there are several interpretations of the Bible, but we know there is only one correct interpretation. How do we know which is true? The correct interpretation is the interpretation that doesn't contradict the rest of the Bible. Currently, in order for people to feel their interpretation is correct, they have to ignore parts of the Bible.

We can't just ignore parts of the Bible when they don't make sense. Who is to say which part is the right part to ignore? Every disagreement on the meaning of the Bible is simply an argument over which parts should be ignored. Clearly, we are in a situation where the only choice for the correct interpretation of the Bible is the non-contradictory one.

This is also true when it comes to the act of modeling God. For instance, God cannot be always completely just and always completely merciful. Otherwise, how would God handle a criminal? If He were always completely just, He would have to punish him. If He were always completely merciful, He would have to not punish him at all.

Someone could say, "There are times when God is merciful." This is true. However, what principle determines when God is merciful? The principle that determines God's mercy is actually a higher principle because it can be completely true all the time. The non-contradictory model for God must contain characteristics of God that are completely true all the time.

One last point before we determine the set of non-contradictory principles that define God the Creator: we have to understand the importance of defining the meaning of words. When people fail to communicate well, they use "semantics" as a cop-out to let themselves disagree.

Actually, there are two possible reasons for the failure in communication. First, both parties are using different definitions for the same word, or second, both parties are using different words for the same definition.

The solution to these conflicts is for both parties to define the words they are using. After all, we all agree on what the Bible says, yet we still disagree. This is because we disagree on what the Bible means. The four principles of Modeletics will resolve every disagreement.

In these books, we are going to use the words in the Bible, but we are going to determine the definitions according to their ability to be non-contradictory. This book is going to cause you to rethink the definition of several words you use on a regular basis. Take time to think through what you mean when you use these words. This shouldn't be a problem unless you believe the Bible intended to contradict itself.

Now, what is the set of non-contradictory principles that define God?

Definition of God the Creator

God is always completely holy. We will see this means that God is always completely righteous and always completely just (we will look more closely at the definitions of these words in the following chapters). To continue our analogy, God lives on the corner of Righteous Street and Justice Avenue. In fact, God is confined by the principles of righteousness and justice. God can't do everything. God can only do what is righteous and just. Every religion believes God is righteous and just. If a religion states God is anything less than this, then its god can't be a Creator (In polytheistic religions, this would only have to apply to one of the gods: their god of creation). If the religion believes anything in addition to this, its concept of God will lead to contradictions.

Let me be clear about this; God has a lot of characteristics (e.g., loving, merciful, compassionate, sanctified, clean, etc.). However, all of these characteristics are a result of being righteous and just. None of these characteristics outweigh righteousness and justice.

Some of the biggest questions people have about apparent contradictions with God occur when they value these characteristics as equal to or higher than righteousness and justice. For example, "If God really loved us, why does He...?" This worldview answers these questions and explains the true non-contradictory meaning of faith, grace, love, truth, etc. by showing how they are a result of being righteous and just.

This is a good place to summarize contradictions and God. A well-known example people use to prove a contradiction exists is to say, "Can God make a rock so heavy He can't lift it?"

First of all, every contradiction is set up from assumptions or premises. What is the assumption or premise this contradiction is based on? God can do everything. The contradiction doesn't exist because God can't do everything.

Even the Bible lists things God can't do. For instance, the Bible says God is eternal. This means God can't cease to exist. When posed like this, it seems obvious that when we present something God can do that is beyond us, it sometimes implies His inability to do the opposite.

The Bible also says God can't lie (Numbers 23:19; Hebrews 6:18). God's nature is holy. He cannot act apart from His nature and be unholy. God can only do what is righteous and just. This leads to another truth about God; God does not have a free will. God can't do just anything.

Free Will

God is righteous and just. God's lack of a free will is not actually a quality of God; it is a result of being righteous and just. This doesn't mean God can't make choices. It means God can't choose to act apart from His nature. The non-contradictory definition of free will is "the ability to act apart from one's nature." Again, God can make choices within His nature. God cannot act apart from His nature and make a choice that isn't righteous and just.

We humans aren't righteous and just, even though everyone aspires to be. This is just one aspect in which we are made in God's image. The quality that makes us human, however, is our free will. We can act apart from our nature when we let God act through us. Free will is the essence of our human ability and the ultimate cause of our mistakes and need to think.

So, here is the model we have for God and humans:

God Model	Human Model
1. Righteous	1. NOT righteous
2. Just	2. NOT just
3. NO free will	3. Free will

If we stop at this point, we can objectively see the complete disconnect we have from God. God is righteous and just; we are not. We have a free will and can act apart from our nature; God cannot.

The only way for God to bridge this gap between His nature and our nature was for there to exist a being that had a free will and chose to be righteous and just in everything he did. This would make this person fully God because he is always completely righteous and always completely just. This would make the person fully man because he has a free will. He could choose to act apart from his nature.

This is how Jesus is able to be fully man and fully God. (Notice, the explanation that Jesus' father is God and His mother is a human would only make Him half man and half God.) This person provides the only way for man to have relationship with God. This is the Jesus Model.

God Model	Jesus Model	Human Model
1. Righteous	1. Righteous	1. NOT righteous
2. Just	2. Just	2. NOT just
3. NO free will	3. Free will	3. Free will

This is why Jesus is the only way to the Father. This is also why Jesus existed from the beginning and through Him, all things were created. He had to be able to knowingly choose to bridge this chasm (because of His free will) before He was physically born into our world. Jesus is the physical embodiment of righteous and just.

God is intentional and lacks contradiction. It is our free will that allows us the option of acting apart from our nature, which is not right or just. Our nature causes us to try to live with apparent contradictions and avoid acting intentionally. We have a free will and can choose to act according to our nature or not; God cannot. Everything comes back to being righteous and just.

God Model
1. Righteous
2. Just
3. NO free will

Some people believe it is impossible to model God. However, a model can express every person's belief in God. Everyone has a model for God whether they realize it or not. Their confusion is due to the difference between their model and the model presented above. Every person who has a problem understanding God essentially believes it is possible for God to be wrong and/or unfair. The key to resolving spiritual conflicts is to fix your model for God.

What is your model for God? Is it possible for God to be wrong and/or unfair?

Furthermore, this model is the key to knowing God. When we say we know someone, we are saying we can, for the most part, tell others what this person believes and how they will react to different situations. In fact, when we are wrong about these areas we say, "I guess I didn't know them as well as I thought."

Likewise, this non-contradictory model for God leads to an ability to better understand what God is thinking and how He will react in different situations. The rest of this book and the worldview are really just explanations and proofs for what God is thinking and how He will react in the most important circumstances.

All of this will truly lead to you being able to know God to a greater extent.

Think back to a time when you believed God was unjust.

- Do you believe it is possible for God to be wrong?
- Do you believe it is possible for God to be unfair?
- How would you resolve the contradiction: God is always completely just and God is always completely merciful?
- How would you answer the question: If it was God's will, why didn't it happen?
- What is your model for God?

Joel Swokowski's *Commentary*

"Modeletics™": The goal of having this term trademarked was never for credit or financial gain. This trademark was done for two reasons:

1. This would show clearly the difference between the information this book presents versus a traditional "Apologetics" style book whose objective is to give as many answers to as many questions as a person can think. Apologetics suffers in that it doesn't get down to the doctrine of a concept, Modeletics does. The word apology means "make a defense." Apologetics is focused on trying to prove yourself right, not necessarily getting to the truth. Christian bookstores have sections dedicated to apologetics.

2. Modeletics is the tool to determine the truth. Having Modeletics trademarked was meant to help make sure the reader/student used all four of the God-given tools *together*. If even one of the tools is not used in an effort to model out a concept, that model will not result in truth. All four must be used together. Do you think Christians want Modeletics or apologetics? Which one would God want us to have?

Modeletics Principle #1: Non-Contradiction

Matthew 6:24 — No man can serve two masters: for either he will hate the one, and love the other; or else he will hold to the one, and despise the other. Ye cannot serve God and mammon.

This is an example of Jesus using the principle of non-contradiction in His premiere teaching, the Sermon on the Mount. Notice, Jesus isn't teaching this principle as if it's new, He's using it in His effort to express truth to the people He's teaching.

I've heard people say that if contradictions don't exist, then why do you keep talking about and exposing contradictions? My response is to merely add one qualifier: "contradictions don't exist, in reality."

God is a set of Non-Contradictory Principles

"God is a set of non-contradictory principles." Those principles are shown to be: always completely righteous and always completely just. This is how and why God is always completely holy. When studying the definition and etymology of the word "holy," you'll find that this word means "of one substance." That one substance is made up of the quality of righteousness and the quantitativeness of justice.

"God is a set of non-contradictory principles." I bring forward this quote again due to the persecution that Lenhart faced from this particular statement. I've had a pastor use this line against this book stating that this is a misrepresentation of who God is. I asked the pastor, "how would you describe God?" and his response was, "love, powerful, perfect, sovereign." My follow-up question was, "aren't those principles?" to which he never responded. Stating that God is a set of non-contradictory principles doesn't mean that Lenhart is saying that God is not a Person, a Father, a Loving King, etc. To accuse Lenhart of such would be bearing false witness against the author and what is written in this book.

A question to really expose a person's view and definition for God is, "Can God grow?" If you say "yes", then you are stating God is less than perfect. If you say "no", then you are saying there is something God can't do. God being always completely righteous and always completely just is the only explanation that proves God is the greatest being in existence. What's better than always completely righteous and always completely just?

When someone is asked the definition of a word they chose to use and they don't define it, they are choosing not to be understood. The reality is, they do this to avoid others finding their contradictions. A lot of bullies use this strategy: get you to answer and focus on your contradictions without ever sharing what they believe is the right answer. Why? Because then you will do to them what they did to you: point out their contradictions.

Free-Will

Free-Will is defined in this book as "the ability to act apart from one's nature." This one's a hard pill for many people to swallow, mainly due to this definition being outside of the traditional way people define this term. The way I help people understand this definition as it relates to God is by showing them, God cannot act apart from His Nature because He has no other nature to act according to. "This then is the message which we have heard of him, and declare unto you, that God is light, and in him is no darkness at all" (1 John 1:5).

There is a huge implication to people saying God can lie but chooses not to, and we will understand this when we learn the last principle of Modeletics.

Notice, this chapter clearly states that God has a lot of characteristics as a result of righteousness and justice, implying that God is many things. However, Lenhart taught that God is defined as always completely righteous and always completely just. It would be wrong for me to say that Lenhart thinks he has God all figured out!

God can also be defined as the First Cause (Creator) and the First Cause would have to be intangible (spiritual). Stephen Hawking would say the Big Bang is the first cause, that "God" is tangible. Then he would say that there must not be a "God" because there must have been something before the Big Bang, before the tangible stuff that was created. Since this is obviously a contradiction, Hawking would conclude that "God" must not exist. The "experts" and scientists who use this train of logic define science by what they can see. Interestingly, these scientists are similar to many of the pastors and "scholars" that Lenhart interacted with over this book. They introduce their own premise, prove their own premise wrong, and use the contradiction within their own beliefs to attempt to prove someone else wrong. Hawking isn't actually proving God doesn't exist. When he uses this train of logic, he's merely showing his inability to determine truth. He has a contradiction in his own beliefs and he's projected that contradiction on everyone else, on the universe, and even on God... or "God."

Jesus

Another area that Lenhart was misunderstood and persecuted in was his description of Jesus. Notice, it clearly states that Jesus is always completely righteous and always completely just and states that this would mean He is fully God. It also states that due to Jesus having a freewill, He was fully man. Lenhart agrees with orthodox Christianity in that Jesus is fully man and fully God. The difference is that Lenhart explains *why* that is in a way most people have never heard. It has led to people accusing Lenhart of preaching against Jesus' divinity even though there's a quote in this book that expressly states, "This is how

Jesus is able to be fully man and fully God." What would it say about my interpretation skills if, after reading that statement, I came to the conclusion that Lenhart presented a version of Jesus that is not fully man or fully God?

Mystery and Puzzle

In the Old Testament, there was doctrinal knowledge and understanding that was kept from humans. Once the Holy Spirit was made available to all on Pentecost, we were given the ability to know and understand this doctrine that was once hidden.

When the New Testament used the word "mystery," it stated we ought to know these mysteries if we are believers. In order to understand this better, allow me to explain the difference between a mystery and a puzzle.

A puzzle begins with a finite number of solutions that decrease as we get more information. Think of a jigsaw puzzle: the more pieces the clearer the picture.

A mystery also begins with a finite number of solutions, however, the solutions increase as we get more information until we get all the information, and then what you're left with is only one possible solution. Think murder mystery, a "whodunit?", where the list of potential murderers appears to grow during the story even while we think we know the identity of the murderer as we get more information. However, as it often goes in murder mysteries, we don't truly know the answer until we get all of the right information.

Christianity is a mystery that has been revealed to believers. It is not a puzzle. The mystery is revealed to us because of the doctrine we've been given access to by the Holy Spirit. The doctrine is the right information we need in order to get the right answers to the hard questions.

The Bible says it's a mystery, which is why we have denominations. We get a good distance away from the wall mosaic, get frustrated, and return to the wall, rationalizing why we can ignore the contradictions in our denomination and using apologetics to help us feel right about it. Basically, we treat the mystery of God like a puzzle and the number of possible explanations divides religious individuals into the denominations we have today. Notice, saying these are possible explanations proves they aren't the truth.

CHAPTER 3

Righteous

ACCORDING TO STRONG'S Concordance (which lists all the words used in the Bible and their Hebrew or Greek definitions), holy (qadash) means several things, including, "clean," "purify," and "sanctify."[2] God is clean, sanctified, and pure. Notice, these several definitions remind us of the party directions before we got the address. In modeling, the definitions have to be specific, especially when we are beginning our model. We will need a definition that incorporates these effects of God being holy. The reason God is holy is because God is right and fair. God is righteous and just.

Righteous means God is "right, correct." Righteous defines God qualitatively. Every other attempt to qualitatively describe God either limits or contradicts righteous.

We instantly see the rub in getting closer to God: How do we intentionally become more righteous? How do we become more right and correct in more situations? Our need to become more righteous is most likely not news to you. However, our focus on intentional righteousness is what makes this discussion different from every other examination of this issue.

The ways we are told to become more righteous are the same as we saw in the first chapter before we got the address. We aren't told what makes God righteous so that we can replicate it in our lives. We are given ambiguous directions that only result in frustration. For example, someone may instruct us to "Read the Bible" or, "Trust God more" or, "Be more like God." When we understand what makes God righteous, then we can know how to intentionally apply the principles in every application we encounter. What makes God righteous?

The best way to answer this question is to look at the opposite result. What are the reasons we are not right or correct in our decisions? Every incorrect decision is actually the result of at least one of two causes: 1) Not having enough of the right data and 2) Not interpreting the data objectively.

God certainly has access to all the data. That's why people say, "God is all-knowing." How many times have you heard someone say, "I didn't know about that" when they were being corrected? One of the ways to become more right is to know more. For example, when we have a crisis, we need to focus on getting more information rather than reacting emotionally. Our righteousness depends on our knowing and thinking more, not less.

The other reason God is righteous is because God is objective. How do we lose our objectivity? We lose objectivity when we focus on proving our point rather than being open to learning more. The first sign of a lack of objectivity is the appearance of apparent contradictions and the individual's response to them. People who have lost their objectivity will fight for the existence of contradictions in their beliefs, even though it proves the belief wrong. In fact, we try to allow contradictions in the areas in which we are comfortable and don't want to change.

Modeletics™ Principle #2: Growth

In reality, everyone is either pursuing growth or comfort. People who pursue comfort want to be right, right now. They want to avoid anything that doesn't make them feel good immediately. In fact, they don't want to have any conflict at all. In groups, their goal is for everyone to get along.

People who pursue growth want to be right in the future. Growth is painful. They are willing to go through short-term pain in return for more understanding. They will even embrace conflict because they know this is the only way to prove truth. When we pursue comfort, we lose our objectivity. When we pursue growth, we gain it.

It's very easy to determine whether a person is pursuing growth or comfort. Watch how they handle a contradiction. The person pursuing comfort will fight for the contradiction to exist, even though it proves they are wrong. In fact, they almost crave the contradiction because it gives him a reason not to think. They may even say, "See, it doesn't make sense. It can't be understood so I don't have to try." We saw this in the first chapter with the people who say they have an "artistic" perspective. They have found an excuse to not have to think about issues that expose their contradictions.

The person pursuing growth sees a contradiction as an opportunity to learn. It is an opportunity to become more right. This person will immediately focus on his assumptions and try to figure out which ones are incorrect. Determining what isn't true is something we can do and know for sure. In fact, as humans, all we know for sure is what isn't right.

Modeletics™ Principle #3: Contrastive Thinking

Let's say I drive a Camry, but I want to convince you I drive a Corvette. All I have to do is describe my vehicle and list all the things that are similar between a Camry and a Corvette. Even though each piece of data you hear makes it more and more likely that I really drive a Corvette, you don't know for sure. But you will instantly know for sure I don't drive a Corvette when I say something that isn't similar between a Camry and a Corvette.

Our brains work from a comparative perspective without any intentional effort. It's in our nature. It is man's way of thinking to look for similarities and consider them as facts. The instinctual force that drives us towards comfort tries to be correct right now (in the short term). In reality, all we can be certain of is what isn't a fact.

If we want to pursue growth instead of comfort, we need to actively prove our beliefs (premises) wrong. After all, the only people who are never wrong are the ones who are perfect or haven't learned anything. I call this perspective "contrastive."

Saint Paul wrote in this contrastive fashion. He would list all of the possible options, then prove one option was right by showing that every other possible option was incorrect. This is what makes his writings so powerful, but also so easy to misunderstand. It is the opposite of the way we naturally think.

A contemporary example of the difference between comparative and contrastive thinking is the Challenger disaster in 1986. The launch group signed off on the launch. Why did they do this? They looked at the data and said everything matched up to a launch scenario. After the shuttle explosion, they looked at the same data and realized it said there would be a disaster. How can the same data show two different scenarios? Isn't this contradictory?

The first time they looked at the data, they weren't objective. They were comparing it to a launch scenario. They only looked at the data that agreed with their goal, the data that made them comfortable.

This happens a lot, especially in groups. One person will state a belief. Other people will sort through all their personal information in order to find facts that support the initial statement. They don't realize that the most valuable information they have is the facts that contradict the initial statement. This may even be due to the leader of the group discouraging disagreement in group settings. As more people share comparative facts, the belief begins to look more right.

After the Challenger disaster, they looked at the data to see if it compared to data that led to an explosion. It did. However, their original objective wasn't an explosion, so they hadn't looked at it from that perspective. They weren't right because they didn't interpret the data objectively.

Summary

Righteousness is one of the characteristics God has all the time. From our previous example, it is a principle that determines God's mercy. Romans 9:15 says that God will have mercy on who He wants to have mercy. This means that God will be right when He has mercy on someone.

If we want to intentionally become more right, we should focus on proving our beliefs wrong. Whatever stands up against that scrutiny is right. Whatever doesn't survive needs to be revised. In fact, contrastive thinking is the solution to world peace. If every person focused on trying to prove his or her personal beliefs wrong, we'd never have another war or disagreement. Yet we live in a world of comfort-minded people, so the overwhelming majority of people are going to have a comparative focus and try to prove themselves right.

We can intentionally increase our ability to be right by learning more and becoming more objective. Again, this requires thinking. This is God's way of thinking; however, we have our own way of thinking. Now we see why God's ways are so much higher than our own (Isaiah 55:8-9). This is not easy to do. It doesn't come naturally because it is not in our nature, so we have to do it intentionally.

God has all the information and can correctly interpret the impact of the data because God is objective. I call this big picture and long term. God understands what will happen in all cases (big picture) over time (long term). Doesn't that set God apart from us? We only see the moment.

For example, in Genesis when Joseph was in jail, was this a good thing or a bad thing? Those who confine themselves to the moment (short term) believe this was a bad thing. Those who look at the big picture and long term realize this was a good thing because it led to Joseph's elevation in Egypt and Israel's salvation.

In order to get closer to God, we need to intentionally grow in righteousness. This means learning more and becoming more objective. We need to pursue growth and remove contradictions from our beliefs. We need to judge every situation with respect to the big picture and long term. This is not in our nature, so we are going to need help.

The only way to become more right is to admit you were wrong. You have to change. "Because they have no changes, therefore they fear not God" (Psalm 55:19).

If you can't admit you were wrong, you are essentially confessing that you are wrong and you aren't interested in becoming more right—and more like God.

- What is the last deeply held belief you have changed?
- How long ago was that?
- During your last life crisis, what did you do to get more information and become more objective?
- Thinking back on your life, would you say you are more growth-driven or comfort-driven?
- How do you respond to others with different beliefs and how do you feel when others are contrastive to your beliefs?

Joel Swokowski's *Commentary*

Qualitatively: Right defines God's Nature characteristically. Right being qualitative is similar to binary: 1 or 0, on or off. It either is something or it is not. Another familiar description from Aristotle would be: A is A.

"God is all-knowing": You may hear this phrase and think "Omniscient." This is often attributed to God as part of how to define Him, or part of His Nature. This would be contradictory, proven out by this question, "If I had all the information available, would I be God?" The obvious answer is "no," made clear due to the fact that I would still be wrong or unfair even if I had all the information. Furthermore, if I'm going to define Omniscience in a non-contradictory fashion it would be "having all the information that exists." I could even say, "God has all the information that is right and just for Him to have."

Modeletics Principle #2: Growth

John 15:2 — Every branch in me that beareth not fruit he taketh away: and every branch that beareth fruit, he purgeth it, that it may bring forth more fruit.

Here again, we have an example of Jesus using the principle of growth. Growth is uncomfortable (purgeth) but that discomfort is a benefit when it results in growth (more fruit).

This principle is the driver behind the other three principles of Modeletics. When a person is growing, the proof is shown by that person using and embracing the other three principles for themselves. This brings up one thing I've seen time and time again, even from people who have read and accepted Modeling God as truth: these principles of Modeletics are meant to be used on ourselves. I'm not growing by showing you where you're wrong. That is a misapplication of the principle of growth. I'm not being contrastive when I point out a wrong belief that you have that doesn't require me to change at all. For example, a person who considers themselves a very right-wing Republican showing a left-wing Democrat where they're wrong is not using the principle of contrastive thinking the right way. The right-wing Republican would have to take a belief of their opponent and consider where they, the right-wing Republican, could be wrong and the left-wing Democrat right. This doesn't mean the person has to believe the other person, but can they consider it?

Modeletics Principle #3: Contrastive Thinking

> "Or how wilt thou say to thy brother, Let me pull out the mote out of thine eye; and, behold, a beam is in thine own eye? Thou hypocrite, first cast out the beam out of thine own eye; and then shalt thou see clearly to cast out the mote out of thy brother's eye" (Matthew 7:4-5).

Again we have Jesus using this principle of contrastive thinking and once again, it's being used in His premiere teaching, the Sermon on the Mount. Jesus taught that before I help a person with their problems, I ought to make sure I'm finding the areas in my own life where I'm wrong and need help.

Contrastive thinking was the hardest of the Four Principles for me to understand. I get the "find where I'm wrong" idea but the "all we know for sure is what isn't right" part confused me. Seriously, I feel like I know for sure when I AM right! What's the deal?

Now I see it plainly with just one clarifier: the reason humans only know for sure what isn't right is because we don't have all the information. It's always possible that I'm missing some vital fact in any of my beliefs. I know for sure when I'm wrong but when I'm right (or more accurately, when I am convinced that I'm right), it's possible I'm not. I may be missing something.

In the time since this book was written, "Growth" is now referred to as "Growth Minded" or having a "Growth Mindset." There is a lot of confusion about what that term means and it turns out that taking these two Modeletics principles together removes the contradiction!

Growth minded is not proven when someone pursues growth. We all eventually pursue growth. We choose when we want to be uncomfortable while pursuing growth by choosing how quickly we admit we're wrong. The sooner a person admits they are wrong, the more they are embracing contrastive thinking, and the more quickly we will see their growth. A growth mindset is measured by how quickly a person admits they are wrong.

Commission vs. Omission

Another concept that was briefly covered in this chapter broaches a bigger topic: Commission vs. Omission. An error of commission is when you do something wrong. An error of Omission is when you don't do something that could have been right.

I've preached an entire sermon on this topic and it took a few weeks afterward for everyone to really "get it." It's hard to hold the perspective that it's okay to do something wrong. It's even harder to hold the perspective that it's better to do something wrong than it is to not make a move at all. We'll see as we get further in this book that the ultimate application of contrastive thinking involves our salvation and the measure for if we're really saved. I'll say this now: every human is wrong. Becoming a Christian does not result in us never being wrong again. This means that even as Christians, especially as Christians, it's our responsibility to find these wrong areas in our lives intentionally and to respond well when we do find these areas.

God cares about your growth. He knows He's Right and Just and He knows you're not. He won't hold you to always being right and just. He will however facilitate in you a right and just response to when you're wrong. What's more important, never being wrong (good luck!) or your growth?

Errors of commission result in learning and growing. Errors of omission are comfort-driven and not immediately obvious. Every organization that falls does so due to errors of omission, not commission. We can also see this expressed throughout scripture.

For example, we've already referred to the following verse from Revelation 3:15: "I know thy works, that thou art neither cold nor hot: I would thou wert cold or hot."

This came from a letter Jesus wrote to the church in Laodicea. Jesus wanted the people there to go **cold** or **hot**. The terms "cold" and "hot" were directional. Go hard one way or the other; state clearly and boldly what you believe. Because God is Right and Just, He will correct you if you are wrong. Being *lukewarm* describes a person who does not state their will, giving God nothing to respond to.

God responds through Justice to a person who is gaining or losing value. This means that God cannot respond to a person who is doing nothing. The individual ought to go towards what they believe is right, believing it is HOT, and then be willing to find out if it was COLD or HOT based on God's response. Let me emphasize this again, I'm not saying this is an excuse to willfully do something wrong. I ought to strive to do the right thing in all situations. Yet I shouldn't worry so much about being right that I never make a decision or take an action. Remember, even if my decision is wrong, God is Just to show me and help me grow!

Another example from the Bible comes from David. He made many errors of commission when he committed adultery with Bathsheba in 2 Samuel 11. However, what was the cause for all of it?

The cause of David committing the sin of adultery was that David didn't go to war with his army when kings are supposed to go to war. This was an error of omission in that he didn't do something that he knew was right. This first error of omission was the cause for all the errors of commission that happened as a result and is missed by nearly everyone.

Hooray, I'm Wrong Again!

Without the ability to identify and correct mistakes, learning how to improve decision-making does not happen. When I do something right, it only confirms what I already know: 1+1=2. Okay, I got it right, but I didn't grow. It only confirmed that I already knew the answer!

A mistake is an indicator of a gap in one's knowledge. Growth takes place when a mistake and its causes are identified, and it is corrected.

CHAPTER 4

Just

"JUST" MEANS THAT God operates according to justice; this defines God quantitatively. Justice says everyone will get exactly what he or she deserves. Everyone will be rewarded for all the good and punished for all the bad. Every other attempt to quantitatively describe God either limits or contradicts His characteristic of being just.

At first, this talk about justice sounds very naïve and childish, but when you think about it, every religion believes in justice. Every religion is based on the idea that you can't do whatever you want without paying for it. Every religion believes in being rewarded for doing good and being punished for doing bad.

The most familiar version of this is the idea of karma. Karma says that when you put out good energy, you will receive good energy. Likewise, when you put out bad energy, you will receive bad energy in return. Let's be clear—this is not fate. Karma does not say there's nothing you can do about your future. It says the opposite. It says you are determining the direction of your future, but not specific events.

We know justice exists intuitively. This knowledge creates guilt when we do something that deserves punishment. If we didn't believe we would have to pay, we would be denying justice. That is the same as denying God's existence. The most familiar application of justice occurs in stories, especially movies. Their plots tend to be driven by the need to resolve an injustice. These stories are universal and consequently, so is the concept of justice.

The Bible references justice when it uses the phrases "with the same measure" (Luke 6:38), "eye for eye" (Exodus 21:24), "for whatsoever a man soweth, that shall he also reap," (Galatians 6:7), etc. Justice is also seen in comments about how someone will receive little recognition after this life if they receive great recognition during this life (Matthew 6:2). This leads to a big-picture view of justice.

The Old Testament has several passages that have a small-picture view of justice. In fact, some would say that the Old Testament speaks about justice not existing. In the Old Testament, Job, Solomon, and David questioned justice. They saw evil people prospering and asked how could this be just—"Yet ye say, The way of the Lord is not equal" (Ezekiel 18:25). When the response was that these evil people are yet to experience a horrible punishment, the writers stated they saw good people die early. How can justice exist without contradiction?

If we take a big-picture view, we realize justice is upheld in the end. This world is not the end. A lot of famous literature misses the boat on this point. God can't violate justice. If God were to be unjust, then everything would cease to exist.

Remember, we have a free will. We can choose to do anything apart from our nature. This is one of the major differences between God and us. We can choose to be holy or unholy; God cannot. God gave us this ability because it would have been unjust to control or limit us. This is why things in this world are not perfect—because God is just.

How can God punish (give bad value) without it being unjust? How does God make sure He is holy?

Justice Applications

God punishes justly by giving a good value and getting a bad value in return. This allows God to settle the accounts by "punishing" the unjust. Look at Moses and Pharaoh for example.

The process began when God said to let His people worship Him in the wilderness (Exodus 5:1). Pharaoh denied Moses' request and made work harder for the Israelites (Exodus 5:4–9). It is important to note that there is nothing wrong with considering or thinking about the request. Today, there are people who think it is a sin or unjust to consider a thought different from their beliefs. They are acting like Pharaoh.

Pharaoh considered the request an injustice that he had to even out. Pharaoh's action to make work harder for the Israelites actually dealt justice against Pharaoh and applied it to God. It is at this point that God tells Moses the rest of the plan. God knows every act He carries out to equal justice will always be seen by Pharaoh as a new injustice that needs to be equaled out.

This allows God to punish Pharaoh by turning the river to blood (Exodus 7:20). Again, at this point everything has been equaled out. However, all this application of justice did was make Pharaoh angrier (Exodus 7:22). The rest of the story is simply a ramping up of justice against Pharaoh.

In fact, Pharaoh made it worse by agreeing several times to let the Israelites go and then backing out each time. Every time Pharaoh backed out of a deal, it allowed God to justly repay with even more damaging plagues (frogs, boils, hail, locusts, etc.), leading eventually to the firstborn being killed during Passover.

The concept of justice runs throughout the Bible and serves as the engine that makes things work. For example, there are three places in the Bible where we are told to love our enemies. Jesus, Solomon, and Paul recommended we do this. Why?

This is done so that we can get a reward and bring justice on our enemies' heads. Jesus said there is no reward from God for being kind to your friends because they will be kind back (Luke 14:12). In fact, they may return greater value to us, and then we are the ones who owe! Being kind to people who won't pay you back is one sure way to get a value from God. God wants us to avoid repaying misdeeds with evil so we can receive a value from God.

Notice how the New Testament never says to take revenge on someone. We are humans and prone to err. If we try to administer our own justice and are wrong, we will be due a greater penalty. For example, Jesus warned us not to pull up the tares because we might accidentally pull up the wheat (Matthew 13:29).

God always tells us to be kind so that He can handle it. Those religions that say it is okay for individuals to resolve a personal injustice are short term and contradict a God that is big picture and long term. Often though, when people are wronged, they do get their justice apart from God through any number of ways. For example, they attack back, take "an eye for an eye," they badmouth the person, etc. If they take any of these actions to an extreme, they may even end up deserving a punishment from God!

Remember—this world is not the end. Ultimately we have a choice whether we want to be paid or punished in this world or the next. If we decide to avoid our punishment here, we will be punished after this life.

Most importantly, justice is at the heart of Christ's death and resurrection. In order for God to provide us value to cover our sins, God had to obtain this value justly. God couldn't give it just because He wanted to or create it out of nothing—that would be unjust. How did God justly acquire the infinite value He would need to pave our way to heaven?

Jesus gave value and never did anything worthy of punishment. In fact, His life was ended without just cause. There is no telling how much value Christ would have given if He were never to have been crucified because we don't know how long He would have lived. Since He is "the life," He could have lived forever. He was owed life and because his was ended without just cause, much like dividing a number by zero, justice said He was owed infinite value. If Christ had sinned, the value He was owed would have been finite. That value may have only been the difference between the good He did and His sin.

God obtained this infinite value by using justice. The blood sacrifice of Jesus on the cross is the source of atonement for our sins. Justice is the source of Jesus' power over death and His ability to give us that value. Later, we will look at this in more detail.

Summary

Justice is a key principle that makes everything clearer when we account for it; however, we often look at the small picture and short term. That is our nature. God looks at the big picture and long term. The more we do things in the big picture and long term (and against our nature), the closer we are getting to God and His perspective.

After all, we are both physical and spiritual beings. Which one lasts forever? Which one should have preeminence? Clearly the physical runs down and is never perfect. In fact, it eventually ceases to exist. The spiritual can be improved and lasts forever. We need to focus on the spiritual in order to overcome short-term, unholy decisions.

If we are focused on intentionally making progress, we will look at the big picture and long term to determine our objectives. That is assuming we want to make progress. Once again, we must determine our goal. Is it growth or comfort? People pursuing comfort want their justice immediately in a physical form. People pursuing growth overlook the short-term physical in order to achieve a long-term spiritual reward.

Let's summarize this with one example of a way God used justice. Jesus gave a good value to the religious leaders through parables. Jesus shared enough of the right information for the hearers to be able to understand and intentionally grow.

It was the listeners' choice to think objectively. If the listeners didn't return a good value, they were without excuse. It was then just for God to take a value from them. God can't judge people if they aren't guilty. How were these people guilty? They chose not to think. People who don't think are testifying against themselves. Again, we see that getting closer to God requires thinking more, not less.

Righteousness and justice are non-contradictory principles. They apply in the big picture and long term. It's pretty obvious that righteousness and justice are intertwined. You can't be right without being just and being just is the right thing to do in the big picture and long term.

Let's look at some applications of righteousness and justice.

- When someone has wronged or hurt you, how have you responded?
- After you responded, how did it make you feel?
- From what you now know about God, how do you think He would have wanted you to respond?
- Do you think it is ever right to retaliate?
- In what instances would it be right to retaliate?

Joel Swokowski's *Commentary*

Quantitatively: "Just" defines God's Nature by measure. "Just" being quantitative can be measured in terms of more and/or less. Another familiar description would be: A = A.

As stated in this chapter, we all intuitively know that justice exists. This is made most clear when someone does something wrong to you. One of the biggest problems in the world is due to people embracing "one-way justice." We think we are owed for the wrong done to us but we don't think we owe for a benefit we get without a just cause.

We've all been guilty of embracing "one-way justice." And there's hope, especially when we take to heart the teachings of Jesus from Matthew 7 that I referred to in the previous chapter. Am I holding you to a higher standard than I hold myself to? If so, that would be unjust of me!

An example of "one-way justice" has been present throughout this book and especially in my commentary is our natural tendency to use the Four Principles of Truth on everyone else while refusing to have them used on ourselves. One way!

How about a pastor accusing the author of a heretical definition while refusing to define the term themselves? One way!

Or even this, a boy using his sister's toy while not sharing his toys with her. One way! Or that same boy stealing a toy from his sister but getting upset when she uses one of his! One way! Sadly, the examples with the children sound all too familiar…and it makes me feel like I'm a child every time I still act that way, even as an adult!

The Mechanism

I was struck very quickly with how the information in this book gave me the answers to the *big* questions. I remember years ago having lunch with some old friends of mine. My friends Steve and Fred are both from Christian homes. Steve and his household continue to this day to serve the Lord. I'm not sure about Fred. Fred was the common story we see of a young Christian man who didn't have a foundation for the beliefs he was taught in church. He went to college and the answers to the *big* questions he was given there made more sense than the answers he was given at church. At our lunch, Fred started questioning my faith. Steve got involved in the discussion and gave the abstract answers that we hear in church still today: "It takes faith," "It's a mystery," "We can't know how this works," etc.

It all came to a head when Fred brought up the September 11 attacks in NYC. I claimed that man was at fault for those decisions. God could not violate the free will of the attackers. Fred was smart, he had thought this through and his response was, "why didn't God break the airplane, flatten a tire, or any of those things He could have done to stop the planes without violating a person's free will?"

Steve answered, "We can't know how these things work, we just need to trust God."

I answered "Justice." The point is, Fred agreed that God does move for or against people in certain situations. The real question he was asking was, "What is the mechanism that determines *if* and *when* God moves?"

Again, I answered, "Justice." God moves for, against, or not at all in *response* to *justice!* This is the mechanism that facilitates God's Will. Although I may not and often do not have the answer to why something is or is not just for God to do, I can rest easy knowing that everything He does is always and completely Just. One of the ways I celebrate the greatness of God is in the recognition of the seemingly infinite amount of variables He's accounting for in every moment. Every action has consequences. God has to and is accounting for them all in every decision He makes, for every person alive. Now *that* is an amazing and *big* God! I may not be able to explain to Fred and Steve the justice involved in the September 11 attacks, but I am confident that God's role (or his lack of a role) in that disaster was a Just role.

CHAPTER 5

Applications

SO FAR WE have taken an intentional approach to understanding how to improve our relationship with God. This approach is based on a modeling technique that actively looks for contradictions in what we believe and removes them.

Therefore, it looks as if our beliefs will change over time. In fact, the only people who don't change their beliefs are the perfect and those pursuing comfort. God allows us free will so we can choose to pursue growth or comfort. Those who are pursuing growth are going to learn more and change what they believe over time. They are going to become more holy, which is not in their nature.

We have found that God is a combination of two non-contradictory principles: righteousness and justice. Righteous defines God qualitatively. Just defines God quantitatively. God does not have a free will. God cannot act apart from His nature. God can only do what is righteous and just. Every other quality God exhibits is derived from His being righteous and just. The rest of this book will show how salvation is a result of a holy God who is righteous and just.

Notice, righteousness and justness are qualities every person inherently understands without being taught. Every person, regardless of philosophical beliefs, strives to be right. Every person, regardless of religious beliefs, becomes upset when they are treated unjustly. Yet, every person knows they are unable to be right and fair all the time because it is not in our nature. In fact, everyone expects others to be right and fair 100% of the time, even though they know it can't be done. These inherent understandings of right and fair are two of the many proofs that God exists.

Before we apply this modeling technique to every other quality God exhibits, we need to understand one more concept: Cause and Effect.

Modeletics™ Principle #4: Causality

The Law of Causality says there is a reason for everything that happens. Things don't happen without a reason. They are not random. We might not understand the reason, but that doesn't mean it doesn't exist. Notice, if everything has a cause, there will eventually have to be an ultimate cause. The First Cause is God. This is another proof that God exists.

Specific causes lead to specific effects, but the same effects don't necessarily come from the same causes. For instance, if I close my eyes for 60 seconds and drive my car 100 mph, the result will likely be an accident. The accident (effect) comes from a person closing his eyes (cause) and driving fast. However, a driver closing his eyes for 60 seconds and driving 100 mph does not necessarily cause every accident.

One of the reasons people view the Bible as contradictory is because they believe effects have only one possible cause. For instance, the Bible clearly states that what is on the inside of people is the cause and their actions are the effect; however, the Bible doesn't say that all people who display similar actions (effects) have the same internal motivation (cause).

As humans, we are very quick to judge people's intentions based on their actions. The Bible tells us in 1 Samuel 16:7 that we judge the outside (effects) but God looks on the heart (cause). That is one of the reasons we make mistakes. We saw in the last chapter how justice will require a payment from us when we are wrong.

Let me be clear. I'm not saying you can't know the causes by looking at the effects. We are supposed to do that. However, immediately assuming the effect has only one possible cause can lead to wrong conclusions. We may need to be contrastive and get more information.

For those who don't believe in cause and effect, there's really no reason for them to continue reading this book. If you don't believe in cause and effect, you can't believe in intentional living. You simply need to wait for things to randomly happen to you. For those who believe in cause and effect, the question quickly becomes, "What are the causes?"

Righteousness and justice are the causes of everything we see. Things can be good depending on how much they align with being righteous and just. The result can be bad by how much they differ from being righteous and just.

The truth is all religions start with righteousness and justice. Any theology they add after this has the opportunity to contradict righteousness and justice. In fact, how a religion links these contradictory beliefs to righteousness and justice is the key to differentiating the religions. Let me be clear about this; while we have used Christian examples, none of the modeling principles we have stated so far are unique to Christianity. Every religion believes in the "God Model." It is the believer who is contradictory.

Examples

In December of 1994, a famous atheist named Antony Flew stated that God must exist. His reason? Causality. He said causality proves there must have been a first cause and that cause would be God. He continued to say that this God could not be the God depicted by the Christians or the Muslims. Why? Non-contradiction. He said these religions depict a contradictory God.

Let's look at another example. The Bible says, "Reprove not a scorner, lest he hate thee: rebuke a wise man, and he will love thee" (Proverbs 9:8).

First of all, the structure for both sentences is cause and effect. However, notice this doesn't mean that everyone who hates you is a scorner and everyone who loves you is a wise man. This is an abuse of causality. Next, when you reprove someone, you are being contrastive and making the person uncomfortable. If they hate you, they are proving they don't want to think like God, and the Bible calls them a "scorner." Finally, the person who appreciates the reproof is someone who wants to remove contradictions. They know your contrastive thinking will cause them to grow in the long term. The Bible calls this person "wise."

Another abuse of causality is sovereignty. Some people believe our lives are predestined and there is nothing we can do. Some even believe God is responsible for everything that happens to us, both good and bad. We will have to wait until later before we deal with this issue completely because it is a complicated belief and it requires more understanding; however, there are a couple of points that should be made now.

First, the Biblical definition of predestinate is "to limit in advance." Certainly, we are all limited in advance as to the things we can do. For example, males are predestined never to birth children. Predestined does not mean our options have been limited all the way down to fate with there being only one possible occurrence for each second of our lives.

Second, this abuse of sovereignty results in a God who intentionally made people for the sole purpose of existing in hell for eternity and being tortured, because there was no possible way for their existence to turn out differently. Not only is this not a loving God, but this belief also presents a God that is unjust. If one were to seriously follow this line of thought to its logical conclusion, the result would be that ultimately everything is God's fault and none of it is ours.

When these people are pushed to explain this, we eventually get the answer we saw in the first chapter and they say, "We can't know God's purpose and just have to find out in heaven." We can conclude the sovereignty discussion by answering what some people like to call "The Hard Question."

The Hard Question

Some people have said the hard question is, "How come God's will doesn't happen?" The thought process goes something like this: if it is God's will for something to happen and God is all-powerful, why doesn't it happen? This deals with sovereignty.

If sovereignty is a cause, then God could unilaterally make His will happen. He could initiate events that overcome the will of the individual. Some people believe God does this. However, this quickly results in several contradictions. First, isn't this unjust of God?

Second, what happens when God's will doesn't happen? If God is able to unilaterally initiate His will, then when it doesn't happen He must be choosing for it not to happen. In effect, it is His will for His will not to happen. That is definitely a contradiction.

We see proof throughout the Bible that God can't unilaterally initiate His will. For example, Jesus says in the Lord's Prayer to pray that God's will is done on earth as it is in heaven. If God were able to unilaterally initiate His will, why would He need us to pray for it?

We've already seen that God can make His will happen in response to justice. It looks as if God's sovereignty is an effect, not a cause. Now we can answer the hard question.

When God expresses His will to someone and they follow it, God is also expressing His will to others that are involved in God's plan. At any time, any one of the individuals involved in God's plan can choose to do something in opposition to God's will. This act of rebellion can hinder God's will for the individual even though the individual did everything right.

The problem is that this situation usually ends up with the individual wondering what they did that was wrong, doubting the ability to hear God, or even doubting the existence of God. Processing this event can take years if the individual doesn't have a non-contradictory model for God.

In the previous chapter, we covered the interaction between Moses and Pharaoh. In Exodus 7:3, God says He will harden Pharaoh's heart. People look at this verse as proof that God unilaterally initiates His will, as if God reached into Pharaoh's heart against his will and there was nothing Pharaoh could do about it. They believe sovereignty is a cause.

Actually, this verse occurs after Pharaoh has expressed his will and initiated an injustice. God is actually telling Moses that He will harden Pharaoh's heart because of, and in response to, Pharaoh's initial expression of his will. God's sovereignty is an effect, not a cause.

Every interaction between God and humans can be explained with justice. Every time God moved on someone's behalf, moved against someone, or refused to move, the reason can be understood by determining where the person stood relative to justice. Unfortunately, people don't have this understanding, so they reference predestination or sovereignty, and that only leads to more contradictions.

Progress

How do you apply what we've learned in Part I in order to make progress toward God? In our party analogy, you would be constantly aware of how close you are getting to Righteous Street and Justice Avenue. Likewise, you must measure everything you do with respect to righteousness and justice. It is a process.

1. **Establish where you are today.**
   Recall from the analogy that the first thing I asked was where you were coming from. The sad truth is that most people don't know where they are coming from. They don't know what they believe. It's no wonder they can't make progress. Once you establish what you believe…

2. **Actively try to prove your personal beliefs wrong by identifying contradictions.**
With respect to the analogy, this would be the same as figuring out if the street you are driving down is getting you closer to the party or further away. That leads to the final step.

3. **Change your beliefs in order to remove the contradiction.**
That's it! That's the process of growing closer to God! However, you may decide you like the street you are driving on even though you know it is taking you further from the party. Even though you are moving and know you are on the wrong street, progress still depends on your actions.

In order to make progress, you must choose to pursue growth instead of comfort. There are plenty of opportunities in this process for people to choose comfort. It is comfort that causes people not to identify their beliefs. It is comfort that causes people not to change their beliefs. Understanding the principles of righteousness and justice helps you objectively identify the beliefs you need to change. Those who don't want to make progress will pursue comfort.

However, pursuing growth does lead to comfort in the long term. Looking back at our analogy, the trip will go more smoothly if you are quick to repair your mistakes. Your journey will also go more smoothly if you understand the details.

The journey would become frustrating if you drove down any road that looked good at the moment regardless of the direction it took you. This is really a focus on short-term comfort and it will only result in a journey filled with frustration in the long term. You should strive for the smoothest journey possible.

Likewise, your journey towards God will go smoother if you make sure your beliefs align themselves with the primary principles of righteousness and justice in a non-contradictory fashion. The rest of this book will focus on some of these secondary principles (faith, grace, salvation, sin, and rewards). Let's look at a couple of simple principles to get us started.

Applications

We've been using mercy as an example. What is the non-contradictory definition of mercy?

Mercy is the postponement of judgment… but it is only a postponement.

Every other definition of mercy leads to at least one contradiction with the Bible.

Justice says God has every right to strike us when we do something wrong. However, justice doesn't determine when God will do it. God wants us to wake up and fix our wrongs in this world, so He doesn't make us pay for it either in this existence or the next. In order to give us the opportunity to make a change, God allows us time to realize it.

Mercy is the act of allowing an interval of time between the bad act and the punishment; however, it can work the other way. If allowing us more time causes us to do worse things, God is actually being merciful by striking us in order to wake us up. God would want all of us to wake up and repair these things without getting punished, but in the long term, we should want to make things right no matter how much short-term pain we receive.

Another example is suffering. When something bad happens to a person, they have to determine the cause. Some would say justice proves that they must have done something wrong and that their suffering is the effect of something they caused. This isn't necessarily true. We know people do suffer from things they didn't cause. How do we resolve the apparent contradiction?

There must be two types of suffering. One suffering is a cause and the other is an effect. For instance, if I do something wrong, I cause myself (and possibly others) suffering. This suffering is an effect or a result. In this case, I could limit my punishment by actively repairing the unholy cause and effects. If, however, I don't cause my suffering, then the suffering occurs for no just reason. Justice says the suffering is a cause and my benefit (effect) is a result of how I handle it. If I handle the suffering well, then I am due a value. If I handle the suffering poorly, then I receive little to no value. Justice not being upheld in this world actually allows us the opportunity to acquire value.

Going back to "The Hard Question," we now see what the result is for someone who has a non-contradictory model for God. The individual would correctly handle the suffering caused by others not following God's will and the result is that they would now be able to obtain an even better situation through justice.

Finally, one of the best places to see the effects of righteousness and justice is in the Sermon on the Mount. The basic structure of the beginning of the sermon is, "Blessed are the [people who have this cause] for they will [get this specific effect because of justice]." Whether it is the merciful getting mercy or the hungry and thirsty getting filled, Jesus showed that justice is the engine that provides rewards to causes (Justice is the other higher principle that determines God's mercy).

This phase of the Sermon ends with Jesus talking about how those who are persecuted for righteousness' sake will go to heaven (Matthew 5:10). He further says that we should rejoice and be exceedingly glad when men persecute us and say all manner of evil falsely against us because of Jesus (Matthew 5:11, 12). Why? Jesus says our reward in heaven will be great. One verse talked about getting to heaven because of righteousness, the other talked about getting rewarded in heaven. Both are resulting from justice being upheld in the long term.

Part I Summary

We have determined a "God Model" that is non-contradictory and shared by all the great religions throughout history.

God Model
1. God is righteous
2. God is just
3. God does not have a free will (He cannot act apart from His nature)

We saw that humans are not righteous or just; however, humans have a free will. We can choose to act apart from our nature and be righteous and just. Humans have no connection with God apart from Jesus because Jesus is fully human and fully God. Jesus can act apart from His nature so everything He did was righteous and just.

We also determined the process needed to make progress on our journey.

Progress Model
1. Establish where you are today
2. Actively try to prove your personal beliefs wrong by identifying contradictions
3. Change your beliefs in order to remove the contradiction

Now that we know where God lives and the principles needed to make the journey, we can intentionally plan out a smooth trip. Likewise, we are equipped to determine how every principle in the Bible is a combination of righteousness and justice by using these modeling principles:

Modeling Principles (Modeletics™)
1. Non-contradiction
2. Growth
3. Contrastive thinking
4. Causality

These principles aren't in our nature as humans. We don't naturally apply these principles to ourselves. We do, however, apply these principles to everyone else. We will not let others contradict themselves, stay comfortable, think comparatively, and appeal to randomness as an excuse for why they were wrong or unfair to us. This point alone proves our hypocrisy. We naturally hold others to a standard we don't require from ourselves.

These principles can instantly find the flaw in every belief system. Throughout history, every error could have been prevented if these four principles were followed completely. However, when these principles are applied to something that has no contradictions, something different occurs. For instance, when these principles are applied to the

Bible, we end up learning the *why*. Not only do we understand *why*, we understand what to do and can explain it to others.

For example, we applied these principles to two areas: mercy and suffering.

Mercy is the time interval between the bad act and God's punishment.
Suffering can be a cause or an effect.

We understand what to do when suffering is an effect:
> We need to determine what we did to cause it and actively work to minimize the damage.

We understand what to do when suffering is a cause:
> We need to handle suffering well so that we can receive the value justice says we deserve.

The only thing that can stop us from making progress towards God is choosing to pursue comfort instead of growth. If you are determined to pursue growth, then it looks like you have everything you need to find the party—but you may have a problem getting in the door.

- Thinking of the last time you suffered, was it a cause or an effect?
- With what you understand now, how would you handle that situation differently?
- From what you've read so far, is there something that is making you uncomfortable relative to one of your deeply held beliefs?
- Can you identify the contradiction?
- What would you have to change in your belief to remove the contradiction?

Joel Swokowski's *Commentary*

We've seen the implications of free will as it relates to God and His Nature. He still has choice, it's just that all of His choices are within the Right and Just Nature. God moves for, against, or not at all in response to justice. God has several options in how He could respond as long as none of them are wrong or exceed justice.

We also learned about The Hard Question: Why doesn't God's Will happen? Now I want to pose: "The Hard Question #2: Why does evil happen?"

So many people want to use the fact that the world is unfair as an example of why God must not exist. First off, think of that for a moment. Atheists actually have the correct definition for God. They state that because things aren't right and just in the world, God must not exist.

Well, I'll argue the contrastive point: the fact that things in this world aren't right and just is because God **does** exist. It's His very nature that results in man's ability to lead an evil life. We are the cause of the evil in this world and to assume that God is allowing it to happen when He could stop it is another way that people are defining God in a contradictory fashion. For instance, the statement "If God cannot stop evil then He is not all-powerful, and if God can prevent evil but does not, then God is not good" comes from the place of either defining God as "all-powerful" or defining God as "all-good." Both of these concepts are effects, neither of them are causes and therefore they are not how God's nature ought to be defined…leaving us with an understanding that this "problem of evil" lies with man, not with God.

Modeletics Principle #4: Causality

Matthew 16:8 — Which when Jesus perceived, he said unto them, O ye of little faith, why reason ye among yourselves, because ye have brought no bread?

Here we see Jesus using the principle of Causality when He confronts His disciples about their lack of faith. He's not telling them not to reason, He's asking why (what cause?) they are reasoning the way they are. Once again we see Jesus use one of the principles of Modeletics, not teaching it as if it's new but using it as if it's always existed.

Sovereignty: often defined as "God is in complete control." We see this contradicts God's nature. The only way to define sovereignty in a non-contradictory fashion is to see it as "doesn't answer to anyone; the ultimate authority." This would be similar to seeing a country as a sovereign nation, meaning that the nation doesn't answer to any other nation. It wouldn't mean that the country was able to do anything or be right in whatever they do.

Predestinate: another great example of this "to limit in advance" definition is when you think about where everyone will end up after this story is complete. Everyone who has ever existed will either end up in the lake of fire (hell) or in the new Jerusalem (heaven). Everyone who has ever lived is limited in advance to those two destinations.

Now that we have the final principle of Modeletics, let's return to whether God can lie or not. We saw "The Bible says what is on the inside of people is the cause and their actions are the effect; however, the Bible doesn't say that all people who display similar actions (effects) have the same internal motivation (cause)."

If God can lie and chooses not to, what causes are in God that would lead to Him lying? This once again brings to light a question posed as an implication of 1 John 1:5: "What dark causes does God have in spite of the Bible saying God is light and in Him there is no darkness at all?"

Notice, when people say God can do anything, even lie, and the previous question is asked, they never answer that question. This is because they would have to state that God has darkness within Him, and that would be a blatant misinterpretation of scripture.

Causality proves that God has to be intangible because the First Cause is intangible. The intangible is always the first cause. For example, an idea of a building is intangible and the first cause of a tangible building.

A lot of critics never made it this far in Modeling God, partially because these four principles prove people's hypocrisy because they use these to judge others but don't allow them to be applied to themselves (One-Way Justice). Furthermore, the critics have also found out this book didn't state what was wrongly claimed about it by the other, more energetic opposers of this book. For instance, let's look at the sovereignty issue again.

God is sovereign, but it is an effect, not a cause. Sovereign means "doesn't answer to anyone." It doesn't mean the person is right or can make anything happen. Hitler's Germany was sovereign. Humans think if they didn't answer to anyone then they could declare themselves right and that would make them right. Right exists whether the person is right or not.

Finally, in Chapter 1, we talked about how Jeremiah showed God wants to be known. We now see that the issue people have for explaining God is due to whether they are talking about God's Nature (cause) or God's

Personality (effects). Check out this animation that extends this topic to a third dimension and explains it in less than 5 minutes: https://youtu.be/KiAo50FtYaA

The foundation has been set and we understand God's identity. Next, we are going to look at the other great source of confusion: salvation.

PART TWO

The Door

INSIDE

Chapter 6: Faith 86

Chapter 7: Faith Examples 96

Chapter 8: Grace 105

Chapter 9: Grace Examples 115

Chapter 10: Uniqueness 128

Chapter 11: Salvation Process 139

Chapter 12: Salvation Implications 156

Chapter 13: Salvation and Rewards 168

Chapter 14: The Way 180

Chapter 15: Discipleship 188

CHAPTER 6

Faith

NO DOUBT THERE are some who think the party analogy, while clever, is inappropriate to their Christian beliefs. Most likely, they see the party as heaven. They have been taught that once they say a prayer confessing their sinfulness and stating Jesus is Lord, they are guaranteed admittance into heaven. Some call this "The Sinner's (or Believer's) Prayer." It looks as if they view this prayer as a "golden key" that will get them through the door to the party.

Furthermore, they see this prayer as the only intentional act needed to make it to heaven. Their admittance cannot be revoked. This is one of the reasons they don't ask for directions to God. In fact, their belief in the "golden key" causes them to have a false sense of security, and they do not try to understand the principles or identify objective measures in order to determine their progress to the party. How is the "golden key" to be used if they aren't anywhere near the door? Doesn't the inability to revoke admittance violate the will of the individual? How exactly does their "golden key" work?

It looks as if these Christians believe everyone makes it to the party, while only those with the key get through the door. How does the

Bible say we get through this "door"? It says, "For by grace are ye saved through faith" (Ephesians 2:8).

This passage in the Bible about how to "get through the door" rests on the definition of two words. Furthermore, the theology of every division of Christianity rests on the definition of these two words, because their theology is the attempt to resolve the apparent contradictions with everything else they believe. It would seem that having a non-contradictory understanding of grace and faith would be the most important focus for Christians since their eternity depends on it. In Part II, we will use the modeling principles from Part I to model salvation.

"Golden Key" Faith

Some people use the term "faith" when they want someone to believe something they can't explain because it doesn't make sense. We see this when we ask people how to get closer to God and the response is "have faith." In fact, many people will defend their lack of thinking by saying faith is something you have when you can't understand something. They may even argue that faith doesn't require action. They can do nothing. Unfortunately, this leads to the "excuse" perspective—people not trying to understand or even asking why. This goes against everything we've said about being intentional.

The "golden key" people act as if faith is a belief in something you don't understand. After all, their focus is on thinking less because they are unable to explain how the "golden key" is non-contradictory. Their definition may even be, "Believing the impossible."

If these were the non-contradictory definitions of faith, how do you build your faith? It would seem the only way is by believing even more impossible and/or senseless things. Now you can see why some religions prove their "faith" by doing senseless things like letting snakes bite them, drinking poison, jumping off of buildings, etc. The devil had this mentality when he tempted Jesus (Matthew 4:5-6). Also, some people think they are growing their faith by buying things they can't afford. In order to model salvation, we need a non-contradictory definition of faith.

What is your Faith Model?

Biblical Faith

The Bible says "Faith is the substance of things hoped for, the evidence of things not seen" (Hebrews 11:1). So there are actually two versions of faith.

1. **Faith is the belief in something you can't see ("the evidence of things not seen").**
   Do you believe in gravity? Have you ever seen it? While you haven't seen gravity (cause), you have seen the effects. Magnetism is another good example. Can you imagine what people thought when they first realized that some rocks attracted or repelled other rocks? How could they explain it? One great scientist, as late as the 1600s, tried to explain it by saying that one object actually sent out waves in the shape of a screw and the other object had holes in it like a nut, so, depending on the shape of the screw and the nut, the objects would either screw themselves closer or farther away.[3]

   We do not see magnetism, yet we believe it exists. Why do we believe magnetism exists? We believe because we see the results of magnetism. The objects being attracted and repelled aren't themselves magnetism. They are not the causes. Electricity is another great example of a cause we believe in, even though all we see are the effects.

Although you see the effects, you have faith in the causes. We can't prove the causes by seeing them, because the causes can't be seen. The causes are eventually proven with the Scientific method, which consists of posing and testing theories with a contrastive perspective.

2. Faith is believing something will happen that hasn't happened yet ("substance of things hoped for").
How do you know if someone believes something will happen? Their actions prove if they think it will happen or not. Notice, faith is proven by actions, not solely by what is said. In fact, true faith is demonstrated by actions and quiet confidence, not yelling and screaming in order to convince yourself or others.

For example, do you think the sun will rise tomorrow? Of course! If you didn't believe it would, you would spend your time making other arrangements. The natural answer to this question is a calm response. How silly would it be for someone to yell, "Oh, I believe the sun will come up tomorrow!" in response to this question?

Experience and Understanding

Why do you believe the sun will rise tomorrow? It's because it has always risen before. What would you think if the sun didn't rise? You'd wonder if the earth had stopped rotating. It seems that faith is based on two things: experience and understanding. In both examples of faith, the first and easiest thing to rely on is experience, because we see the results of the cause. This is called "experiential faith."

When it comes to the faith in the unseen, we see the results when we run tests. When it comes to faith in a future event, we look to the past to rely on what we have seen because the hoped-for event hasn't happened yet. Experience is powerful, but it is shallow.

If something happens ten times in a row, how confident would you be that it would happen the eleventh time? Probably close to 100%. Let's say it fails to happen. How confident are you that it will happen on the twelfth occurrence? Probably a lot less than 100%. While experiential faith is wonderful, it can disappear in an instant. Thankfully, we can have faith based on understanding.

With faith in the unseen (e.g., magnetism and gravity), we see that faith is built through a theory based on the results of a series of experiments; that is, through trying to understand it. When we can't see the causes, we learn to control them by setting up many situations and determining the results. We are finding the boundaries. This process is intentional and is focused on gaining understanding.

With faith in the "yet to happen," we believe because we understand what causes the expected results. For example, we understand the sun rises because the earth revolves. We are confident that the sun will rise because we have understanding as to why it does.

This faith is harder to shake when something goes differently than planned. Instead of losing faith, we realize that we have probably exceeded a boundary or happened upon a different condition. Rather than feeling helpless and wondering if everything is going to go back to how we expect, our response would be to understand what was different. With the understanding of both applications of faith (unseen and future events), we are focused on understanding more, not less.

Jesus said: "blessed are those who haven't seen and yet believe" (John 20:29). Why? The disciples initially believed in Jesus because they saw Him. Their belief was based on experiential faith. The gospels show they didn't have faith based on understanding. That's why they panicked and lost their faith when Jesus was crucified.

We are blessed when our faith is based on understanding. Even though we can increase our faith experientially, our faith in Jesus should begin with understanding. In fact, our understanding makes us look at the disciples and wonder how they could have lost hope so quickly. We may even say, "Why didn't they understand what Jesus was doing?"

Knowledge, Understanding, and Wisdom

Realize that understanding is different from knowledge and wisdom. The Bible says, "Wisdom is the principal thing; therefore get wisdom; and with all thy getting get understanding" (Proverbs 4:7). However the Bible also says, "Knowledge puffeth up, but charity edifieth" (1 Corinthians 8:1). Knowledge is the ability to know facts. Someone who has a lot of knowledge knows a high number of facts. An example is to know how a motorcycle works.

Understanding is the ability to take facts (knowledge) from more than one area and fit them together to create new knowledge. Understanding *creates* knowledge. It is like modeling because it identifies the implications and the likelihood of possible occurrences.

For instance, if I have knowledge of friction, the physics of bodies in motion, gravity, and weather effects in addition to the knowledge of how a motorcycle works, I can make a prediction. For example, if I drive a motorcycle at a specific speed over a specific ramp with a specific angle, I could calculate the distance of a second ramp that would cushion my return back to solid ground. Notice, this distance is a new fact (knowledge) that resulted from understanding.

Wisdom is shown in our decisions. Wisdom (like faith) is proven by our actions, which are a result of our decisions. We will learn how to objectively measure wisdom in the second book. For now, let's just say wisdom would ask, "Why are you doing this?" Most likely, wisdom would be shown when I didn't jump the motorcycle off the ramp.

Actually, understanding would tell us that there are several things that could cause our calculations to be off. Understanding would even anticipate things that seem to have nothing to do with our model. For instance, what if there was a gust of wind or the ramp gets dirty or slippery from condensation? Understanding is necessary for us to make wise decisions.

We saw the Bible says wisdom is the principal thing. In fact, it links wisdom and understanding. Now we see why wisdom is principal and understanding follows closely behind. We are supposed to make good decisions that are proven in our actions. It is difficult to have wisdom when we just know facts but don't understand the *why* and/or their implications. Likewise, faith is built on understanding and not simply knowledge.

Summary

We've seen that faith is active. Faith is a cause, and its results are seen in our actions. One of the directions we were given in order to get closer to God was "to have faith." While faith is a cause, in order to intentionally have faith, we must look at faith as an effect. Faith is the effect of what causes? In other words, how would you have (or grow in) faith?

Faith is a result (effect) of our experience and understanding. This means increasing your experiences with God and growing in your understanding through the Bible would intentionally lead to growing faith in God. It is possible for you to have more faith in the next five minutes. How? Get more experience with or understanding of the object of your faith.

Understanding and experience cause faith, and faith is one cause of salvation.

Faith Model
1. A belief in something that can't be seen
2. A belief that something will happen that hasn't happened yet
3. Built through experience and/or understanding
4. Proven by actions

Faith should be able to answer the question: "Why?" The reason we read our Bible is to build our faith through understanding. We will look at some Biblical passages on faith in the next chapter.

- What is an area you have strong faith in?
- What understanding and/or experiences led you to have this stronger faith?
- What is an area you believe you have little faith in?

- What experience or understanding would you need in order to increase your faith in this area?
- What would your actions look like if you had greater faith in this area?

Joel Swokowski's Commentary

Presenting the "Golden Key" definition of faith must have confronted many pastors, possibly due to their being guilty of teaching and applying faith in a contradictory way. Instead of being grateful to this book for showing them an area they needed to grow in, they attacked the author. Pastors couldn't answer how to grow in faith, and they proved they didn't want to grow or admit they were wrong, so they attacked the author. This was the first sign of what he was going to deal with as the questions they couldn't answer became more crucial.

The definition presented in this book was hard for these pastors to refute, especially since the definition came directly from the scriptures. This may give you some insight into how stubborn many of these pastors were since their attacks continued.

One piece of proof that many Christians, including pastors, don't understand faith is that they have to qualify it. For instance, look at the expression "saving faith." Why would a person have to qualify the term faith as "saving faith" if they understood what faith meant at the doctrinal level? What faith in God would *not* lead to salvation?

This chapter also presented and defined the terms knowledge, understanding, and wisdom. Knowledge being "facts," understanding the "reason/cause," and wisdom being "a profitable decision." Another way we'll see this throughout the rest of this commentary is:

- Knowledge: right *what* (facts)
- Understanding: right *why* (contextual tangible causes)
- Wisdom: right *how* (doctrine: Spiritual Causes)

This is also a great chapter for proving the importance of using one of the principles of Modeletics. James 2:14-26 gives an in-depth presentation about faith, specifically stating that faith without works is dead. These scriptures have caused many debates in the history of the church. However, when I embrace the principle of Causality, it becomes clear. Faith is the cause, works are the effect. I focus on the subject of my faith and my faith is proven out when the effects (works) are made manifest. I don't need to focus on the works if my faith is true. Without an understanding or a correct application of Causality, it becomes clear when people have an issue understanding these verses.

CHAPTER 7

Faith Examples

OUR DEFINITION OF faith came from Hebrews 11:1. The sixth verse in that chapter summarizes the topic by explaining why faith is so important.

> "But without faith it is impossible to please him: for he that cometh to God must believe that he is, and that he is a rewarder of them that diligently seek him" (Hebrews 11:6).

Again, faith is proven by results (pleasing God) that line up with the definition. It takes faith to believe that God exists because He is something you can't see. It takes faith to believe God rewards those after they diligently seek Him because it hasn't happened yet. Our non-contradictory definition and model are consistent with this passage.

In fact, Hebrews 11 lists what has been called, "The Faith Hall of Fame." The author gives a list of people who demonstrated great faith. If you were to go through this list, you would find every one of the people listed carried out their actions based on understanding and/or experience. None of these people believed something that didn't make sense or was impossible simply because God said it. Whether it was Gideon asking for signs from God or David recalling how he killed animals, each person believed because of understanding and/or experience. Think about that. Gideon is listed as an example we should follow, and he needed signs in order to believe. Abraham asked God for a sign so he could know what God said was true. Furthermore, the author even explains why members of the list had great faith. For example, look at what is written about Abraham.

> "By faith Abraham, when he was tried, offered up Isaac; and he that had received the promises offered up his only begotten son, of whom it was said, That in Isaac shall thy seed be called: Accounting that God was able to raise him up, even from the dead; from whence also he received him in a figure" (Hebrews 11:17–19).

Notice the beginning proves Abraham had great faith because of his actions—he offered up Isaac. The rest of the passage gives the reason why. Abraham understood God's promise that Isaac was the source of his descendants. Abraham also had experience with God that convinced him God would be able to raise Isaac from the dead if that's what it would take to keep the promise.

There are numerous passages that speak about faith. Since we will look at just a few examples in this chapter, it is best to stick to the source. These examples come right from Jesus. The next two examples deal with people having little faith. The final example deals with someone Jesus identified as having great faith.

Little Faith

> "And I brought him to thy disciples, and they could not cure him. Then Jesus answered and said, O faithless and perverse

generation, how long shall I be with you? [H]ow long shall I suffer you? [B]ring him hither to me" (Matthew 17:16, 17).

This passage shows that faith is the cause and is proven by results and actions. Jesus called them "faithless" because they couldn't accomplish what He taught them to do, which was cure people. Jesus assigned no value to what the disciples said. Also, notice how this has nothing to do with believing something and doing nothing.

If the disciples had "golden key" faith, they could have responded that they did have faith because they really believed the person was going to be cured even though it didn't make sense to them and they couldn't explain it. Notice, the disciples didn't disagree with Jesus on this point.

Believing is important, but it doesn't prove faith; results do. Furthermore, Jesus says that perverseness and lack of faith go hand in hand. Having faith and building it is very important. We are told to do this intentionally, especially since it is one cause of "salvation."

> "Then Jesus said unto them, Take heed and beware of the leaven of the Pharisees and of the Sadducees. And they reasoned among themselves, saying, It is because we have taken no bread. Which when Jesus perceived, he said unto them, O ye of little faith, why reason ye among yourselves, because ye have brought no bread? Do ye not understand, neither remember the five loaves of the five thousand, and how many baskets ye took up? Neither the seven loaves of the four thousand, and how many baskets ye took up? How is it that ye do not understand that I spake it not to you concerning bread, that ye should beware of the leaven of the Pharisees and of the Sadducees?" (Matthew 16:6–11).

Jesus got more specific in this passage about "little faith." Jesus said the disciples had little faith because they focused on the physical and worried. Jesus believed they should have known better than to worry about bread. Notice Jesus said, "...why reason..." Jesus is not saying they shouldn't reason. His point is that they reasoned incorrectly. Otherwise, Jesus wouldn't have said, "Do ye not understand."

Besides, if they weren't supposed to think, then Jesus wouldn't have taken the time to explain it. In fact, Jesus doesn't seem to care what they said. His focus is on their thought process; that is, why they said it. Clearly, Jesus wants them to increase their faith, and He's going to show them why they have little faith. Jesus gave two reasons.

First, He focused on faith based on understanding when He asked them why they didn't understand. As stated previously, Jesus asks why they reasoned among themselves about the wrong thing. Essentially, Jesus said they shouldn't have worried about running out of bread because they should have understood Jesus wasn't speaking about the physical. I believe He spoke about faith based on understanding first because that is where they were lacking the most and that is the faith that will last the longest. It is the faith we should be focusing on.

Secondly, He spoke of experiential faith when He asked them if they remembered the two miracles He performed when He provided bread. Jesus said they shouldn't have worried about bread, because they had personally seen Jesus perform two miracles concerning bread. Their experiential faith should have told them they had access to an unlimited supply of bread. They should have realized Jesus could feed twelve if He previously proved that He could feed thousands. The teaching session didn't stop there.

Jesus finishes by asking them how they didn't understand that He wasn't talking about bread. Jesus actually is focused on their thought process. He is interested in helping them improve their reasoning process—and it all comes from a discussion about faith.

Great Faith

> "The centurion answered and said, Lord, I am not worthy that thou shouldest come under my roof: but speak the word only and my servant shall be healed. For I am a man under authority, having soldiers under me: and I say to this man, Go, and he goeth; and to another, Come, and he cometh; and to my servant, Do this, and he doeth it. When Jesus heard it, he marvelled, and said to them that followed, Verily I say unto you, I have not found so great faith, no not in Israel" (Matthew 8:8–10).

Jesus actually marveled at what the centurion said! Can you imagine making Jesus marvel at you? Jesus said the centurion had great faith. Why? What did the centurion say? He said that he believed Jesus could heal the person by just saying the word. What did the centurion do? He prevented Jesus from coming under his roof. Faith was shown by his actions. Let's look at how the centurion demonstrated both forms of faith.

The centurion understood why Jesus could just say the word because he knew how it worked in his life. In fact, he quoted both experience and understanding! Everything after the "For" is the centurion's reason why he believed Jesus could heal the man only by saying the word. Basically, it was experience in his relations with people and understanding of Jesus in the spirit realm.

The centurion understood from experience that anything that functions properly must have order. He understood his place. He stated this twice. He understood he was under authority and acted as the authority in his professional life. He knew from his professional experience the process of authority.

He had understanding of who Jesus was and that the Spiritual realm had to work the same way. This caused Jesus to marvel because the centurion applied this professional experience to his personal life and let his actions follow. The centurion understood he was also under authority in his personal life when he called Jesus "Lord" and said he wasn't worthy to have Jesus in his house. He even supported the actions he was taking with this reason. From the previous faith example, we can even suspect that Jesus marveled at the centurion's thought process.

If the centurion really didn't believe what he said, he would have had Jesus come to his house just in case the servant wasn't healed. The centurion proved his faith by barring Jesus from the house because if his servant wasn't healed, the centurion had no other recourse. His actions showed he truly believed a word from Jesus was sufficient.

Summary

Great faith is proven when a person's actions follow his experience and understanding to believe in something that hasn't happened yet and/or can't be seen. Jesus clearly told us to build our faith intentionally. Jesus measured a person's faith by the results—by his actions.

Faith can't simply be believing the impossible without explanation or understanding. Jesus wouldn't have reasoned in each of these examples if it were not understandable. Hebrews 11 lists people who had great faith. All of these examples involve understanding and experience.

God starts us out with experiential faith. However, if we aren't focusing on backing this up with understanding and turning it into faith based on understanding, then we are going to lose our faith as soon as something happens differently than what we have previously experienced.

- What specific experiences and understanding did Abraham have? (Hebrews 11:19, Genesis 22:5–8)
- What specific experiences and understanding did Gideon have? (Judges 6, 7)
- What specific experiences and understanding did David have? (I Samuel 17:26–58)
- What specific experiences and understanding did Adam and Eve have? (Genesis 2, 3)
- On all of the above, how was faith proven or not proven in their actions?

Joel Swokowski's *Commentary*

In seminary, one of my favorite topics to tackle was "faith." I was able to write a research paper, specifically covering the battle of "Faith vs. Reason." My conclusion was that Faith and Reason are complementary to each other, not contradictory.

You can see from what we've learned about faith and the principles of Modeletics presented earlier that thinking (reasoning) is important to our belief system, to God, and to Jesus. The book's explanation of faith and the way the Bible actually teaches it has caused Atheists to love this chapter. I've experienced many of them saying, "If someone had explained faith like this to me, I would have believed/accepted it". Atheists' main argument against God's existence usually has to do with the explanations they've heard being unreasonable and contradictory. They simply did not "make sense!"

The entire section regarding "Little Faith" proved that Jesus wanted His disciples to think, and to think in the correct manner.

A tremendously popular book had been out for over ten years when Modeling God was published. The book encouraged people to deal with their issues by not thinking. It even misquoted a passage of the Bible to infer that reasoning leads to confusion.

We've already seen that Matthew 16:8 is accurately stated as, "Which when Jesus perceived, he said unto them, O ye of little faith, why reason ye among yourselves, because ye have brought no bread?"

The popular book misquoted this verse as, "…O ye of little faith, why reason ye among yourselves?"

Think about this for a moment… the author of that book changed the scriptures in order to support their point that we shouldn't be thinking! Isn't this an example of massive abuse happening in the church?

Towards the end of this chapter we see the statement, "God starts us out with experiential faith." This reminds me of faith as it relates to the born-again experience (which we'll get into more specifically during the Salvation chapter). The born-again experience is a very emotional and transformational moment for everyone who goes through it, and rightly so. If you're a Christian, you know for yourself and you've seen new Christians seem to have a youthful zeal that is attractive, often eliciting envy in the more seasoned believer. There comes a problem though, not long after conversion if the person who's become born-again doesn't add understanding to their foundation, to their newly adopted Christian worldview. If something challenges their faith and that faith is mainly experiential, it's likely that this person could fall away. If that person is given a strong foundation of understanding, it can help the person deal with the challenges that come their way, specifically the "bad" experiences, and the person can remain strong in their faith regardless of the context.

What have you done to ensure you can handle the "bad" experiences that come your way? Does your faith remain strong or falter during those hard times?

CHAPTER 8

Grace

"FOR BY GRACE ye are saved through faith" (Ephesians 2:8). We have modeled faith. The second definition and model we need in order to model salvation is grace.

When you ask most people their definition of grace, their answer is "unmerited favor." Some people think this was the definition in the Old Testament. According to Strong's Concordance, the word for grace in the Old Testament was "chen," and has a reference number of 2580.[4] It means "kindness" or "favor." It is from the root word "chawan" (reference number 2603)[5], meaning "to stoop in kindness to an inferior." The definition of "unmerited favor" does not come from the Old Testament. We will see that it is based on tradition. Others may even say it is "assigned advantage." If we take a step back and look at this, we have to ask, "Why does God prefer some people over others?" Is this right? Is this just? Is it right and just for some to get value they didn't deserve?

Scarier still are the implications of this "golden key" grace definition as it relates to justice. When we do wrong, justice demands that we pay. However, this definition of grace says we have unmerited favor (value we didn't earn). How does grace work according to this definition?

When a person does wrong, they have to pay. This is the foundation of the principle of justice. This "golden key" definition says the person pays (and continues to pay) with the unmerited value they received from God. Now that their debt is paid and they are righteous, they can go to heaven. This actually raises a lot of questions.

"Golden Key" Implications

If God's grace is boundless, a person could continue to do wrong and never have to worry about running out of unmerited value. In fact, if we are supposed to grow in grace, how do we do that? The way to grow in grace is to sin more so that this "unmerited favor" would flow through the individual more. It seems this definition encourages people to sin more so they can grow in grace. Paul realized this was not the definition of grace and stated that people will misunderstand it in this fashion (Romans 6:1).

Another issue is that justice says we are still going to have to pay for this value we got from God. God may give us value to pay for sins here, but we still need to reimburse God. How would we be able to get to heaven when we owe a huge debt? This definition of grace creates a Grace Model and a Salvation Model that are full of contradictions. What is your Grace Model?

Obviously, this "golden key" definition of grace makes no sense, but that's where "golden key" faith comes in—believing in something you can't understand. So, we can see why people have this model of faith and grace. They can continue to sin because they have unlimited value and it doesn't make sense. Further, they look to the fact this doesn't make sense to prove their definition of faith. When we extend this model, we find the belief (faith) is based on a feeling.

Finally, if grace depends on feelings instead of reason, God would have no issue with people who believe something different than His plan because they just as easily could have felt their contradictory plan was right. God can only condemn people if they have done something opposite of what they know to be right. If faith doesn't take into account understanding but goes by feelings, then God can't justly condemn anyone. It only takes a couple of questions to expose the flaws in this model. What is your Salvation Model?

Actually, the world lives by "unmerited favor." People who aren't Christians are trying to be good. They know they aren't good and don't deserve to go to heaven. How do they live with this knowledge? They hope that God will look on their good intentions and give them a value they know they don't deserve.

If "unmerited favor" is the key to salvation, there's no one who is responsible for being sent to hell. The individual's ability to receive unmerited favor is completely up to God. Basically, salvation becomes all God's fault and if God is never wrong or unfair, no one is going to hell. Unmerited favor is the reason we are seeing Christian writers question the existence of hell.

Yet these people tell us to believe in their God. Why? There is no benefit or penalty in not believing. In fact, belief in "unmerited favor" is the reason studies have shown there is no difference in the divorce rate, the incidence of unwed pregnancy, drug use, etc. between people who do and don't attend church. Both groups have the same beliefs!

Basically, we have to ignore or modify our concept of justice in order to make these Faith and Grace Models consistent with each other and salvation. However we know that: 1) There must be a non-contradictory Grace Model, and 2) Faith is based on understanding and experience.

New Testament Grace

We've seen that "golden key" faith is flawed. God tells us to intentionally know and understand. Jesus focused on people's reasoning process when He spoke of faith. The Biblical definition of faith doesn't mesh with the definition of grace as "unmerited favor."

God wants us to grow in faith and grace, so we must do this intentionally. We've seen that this is going to take understanding. In fact, it is dependent on our reasoning process. As humans, we have a volitional consciousness. We can choose to think or we can choose to not think. This becomes an expression of our will. "Unmerited favor" is a traditional definition. What is the New Testament definition of grace?

Unlike faith, there is no passage in the New Testament that gives the definition of grace. When we don't have a specific definition, we will begin our process by considering another source and then removing the contradictions until the definition agrees with the Bible. In this case, we will start with Strong's Concordance (Greek word and reference in parenthesis).

Grace: The divine influence upon the heart, and its reflection in the life ("Charis" 5485).[6]

There are two steps to grace! First, we have to intentionally let God influence our hearts. We have to focus on strengthening our ability to hear God and our ability to let God influence our hearts. These are actions that are in our control and can be improved with experience.

Further, these are actions that are consistently mentioned in the Bible so they can be better understood. These are actions we could be condemned for *not* doing. Finally, notice this divine influence occurs individually on our hearts. Grace is a process that occurs between God and the individual—there are no other participants.

This first part of grace could be abstractly described as "unmerited favor." God's influence is definitely "favor." Also, it is "unmerited" in that we did nothing to deserve this influence. In all actuality, God is speaking to everyone, whether they are a Christian or not. God's grace extends to everyone. It is our fault if we don't listen to God. However, if we stop here, we miss the real impact of grace. Unfortunately, tradition has stripped Christianity of its full power by recognizing only half of the definition of grace.

The second part of grace concerns our actions. We have to intentionally choose to let this influence come out in our actions; otherwise, it didn't really influence our hearts. Grace is consistent with faith in that it doesn't exist unless it comes out in actions. So grace, like faith, is the cause, and the actions are the result. Grace, like faith, is a result that can be increased with understanding and experience.

Allowing God to influence our hearts and having it reflect in our lives are actions we can control. These are attributes we can increase intentionally. These are actions on which we can be judged objectively. All of this is consistently written about in the Bible. In fact, we are told we will be rewarded for these actions because of justice.

It would seem that grace is actually the best way to improve our relationship with God. What could be better than listening to God and intentionally choosing to let God direct your actions? Yet, while people tell us to, "Have faith," you don't hear them say, "Have grace." Why? That's because when most people speak of grace, they are talking about "unmerited favor." It wouldn't make sense to tell people to "have unmerited favor." Further, that definition puts the responsibility on God instead of us. Since grace is half of salvation, that would mean we don't have control over our salvation. This contradicts free will.

Notice the only way for us to "do the right thing" is for God to do it through us, which is grace. Grace is the ultimate way for us to become righteous because it is God who is working through us. We can act apart from our unholy nature by choosing to let God work through us. Our free will allows us to become holy by choosing to let God work through us instead of walking in the flesh. Righteousness is the key to salvation. "For I say unto you, That except your righteousness shall exceed the righteousness of the scribes and Pharisees, ye shall in no case enter into the kingdom of heaven" (Matthew 5:20).

Theoretically, unmerited favor only removes our unrighteousness. How does unmerited favor result in enough righteousness to obtain salvation? Unmerited favor is not a cause of righteousness and therefore has no power to save us. The only way to have enough righteousness to obtain salvation is to let God work through you. This is the same as the non-contradictory definition of grace.

Summary

We now have a non-contradictory model for grace and faith. Notice how these definitions of faith and grace mesh perfectly. They both depend on an intentional choice to understand and act in accordance with this understanding. They both grow through experience and understanding.

Grace Model
1. Divine influence upon the heart
2. Reflection in the life
3. Grace is built through experience and/or understanding
4. Grace is proven by actions

The "golden key" definitions of faith and grace try to cover the faults of each other. The definition of grace contradicts justice. The definition of faith tries to prevent the individual from thinking about the contradictions caused by the definition of grace. The definition of faith tries to negate reasoning, whereas Jesus used the importance of reasoning to show who had faith and who didn't.

Clearly, the "golden key" definitions of faith and grace don't line up with the Biblical definitions. They are contradictory. The Biblical models are non-contradictory. What does our Salvation Model look like so far?

We gain salvation by how much we individually let a God we can't see (faith) determine our actions (grace) in hopes of a result that hasn't happened yet (faith). The result is righteousness. There are numerous examples in the Bible of grace and actions.

In fact, in Luke 2:40 while talking about Jesus, the Bible says, "And the child grew, and waxed strong in spirit, filled with wisdom: and the grace of God was upon him." Jesus never sinned. Jesus didn't need unmerited favor. In fact, wasn't the favor that Jesus received from God "merited"? After all, He is the Son of God. How can anyone look at this passage and say that grace is "unmerited favor"? If grace is "the divine influence upon the heart and its reflection in the life," this verse makes sense. This definition is the only non-contradictory definition that can be applied to every New Testament use of grace (charis) in scripture. In order to believe in "unmerited favor," one has to accept contradictions, think comparatively, and pursue comfort.

In fact, newer translations of the Bible prove that "unmerited favor" is contradictory. In the most contradictory verses containing charis, the word grace has been replaced (e.g., "mercy") in an attempt to remove the contradiction.

Let's look at some more examples of grace in the next chapter.

- When was the last time you believed you heard God give direction to your life?
- What makes you believe it was God?
- What did the reflection in your life look like?
- What understanding and experience would you need in order to grow in grace?
- Do you see that grace is the only way to act apart from your nature?

Joel Swokowski's *Commentary*

Keynote, the Old Testament word for grace coming from the Hebrew word chen, although translated into the English word "grace", should not be seen as the "grace" of our salvation. I ought to treat the term "chen" and the term "charis" as different words altogether, as I would with two different English words. Having two different words (Hebrew and Greek respectively) translated into the same English word is a weakness in the English language, not a weakness in the original scriptural documents.

Just as "golden key" grace and "golden key" faith work together, so does what could be seen as a "golden key" definition of God: God = love. Yes, the Bible states that "God is love," yet this is not an accurate explanation of God's Nature. We'll cover the concept and definition of love more clearly in a later chapter, but for now, as it relates to grace and unmerited favor, if grace is my "golden key" to going to heaven due to unmerited favor and faith is my belief in something that doesn't make sense, eventually, this all comes to a head by defining God as "love." If God is love and grace is unmerited favor, then the logical progression of that would result in everyone getting to heaven. It's these flawed definitions that lead to man-made religions and belief systems, like "Universalism."

We saw with faith that a red flag to indicate that a person is not understanding it at the "bottom rung" or doctrinal level, is when they qualify it as "saving faith." I've seen this same thing happen with grace where people use the qualifier: "saving grace." What "grace" doesn't lead to salvation? People tend towards needing to qualify words like this when they have multiple definitions for a term based on its usage. Having different definitions based on the context also shows that the person is not at the doctrinal level for that term.

Notice how this chapter derives its definition of grace from the Strong's Concordance. Why then would pastors continue to claim that this was a definition that Lenhart came up with himself? Is this yet another way that the pastors are showing they're more concerned with not being wrong than they are being assured of their salvation and helping their congregation be assured of salvation? Is this once again showing a huge lack of ability in many "experts" and "scholars" to interpret text?

What's the process you use to determine a definition of a word? What Lenhart has done, and what every Biblical scholar ought to do, is to first go to the scriptures for a definition. This worked great with the definition of faith, as it was clearly stated in Hebrews 11:1. With grace, we see another step to take. Since this term isn't explicitly defined within the scriptures, we go next to a concordance, as with the definition for grace presented here. However, the ultimate measure for what the accurate definition is for any term is the principle of Non-Contradiction. This is true whether I find a definition in *what* the Bible says or I find a definition in dictionaries (which are made by men), or I hear a definition from another teacher or book. The definition must be Non-Contradictory for it to be at the "bottom rung" or "doctrine" level.

If we take the concept of unmerited favor and apply it, we see that unmerited favor would occur after sin. This chapter even states, "when a person does wrong, they have to pay." Unmerited favor would only come into play after the person "does wrong." The Biblical definition of grace would occur prior to sin and give us the power/ability to avoid sin. Some people have this definition of grace, which is still one rung up on the ladder of abstraction but isn't contradictory. This means that yes, grace can be described as unmerited favor. I didn't do anything to earn it. Yet grace cannot be defined as such, because unmerited favor is not the cause of grace.

The issue is that people can't separate out responsibility and credit. It looks as if everyone either believes God gets all the credit and is responsible for their salvation, or the individual gets all the credit and is responsible for their salvation. Why do we lump these two when it comes to God?

If I was sick and took medicine, what would you think if I said, "I healed myself"? You might say, "The medicine healed you." However, I was responsible for *choosing* to take the medicine. The credit for the healing goes to the medicine.

Likewise, each of us is sick with sin and we are ultimately responsible for choosing to take the medicine of grace so that God gets the credit. The medicine is a free gift (unmerited favor) that none of us created and we are judged on our decision (responsibility) to take the medicine or not. Furthermore, the medicine can be taken to prevent sickness!

CHAPTER 9

Grace Examples

BEFORE WE LOOK at the examples, consider the following questions:

Can we give grace to others?
Can we frustrate grace?
Can we understand grace?

If grace is "unmerited favor from God," then the answer to all of these questions is no.

We can't give unmerited favor or assigned advantage to others; it is a gift that comes from God. There is no way to frustrate this grace because it comes from outside of us and is greater than us. In fact, it is infinite. The "golden key" Grace Model says that God chooses who He wants to give unmerited value to and it is beyond our understanding. Ultimately, this model would say God has a free will and doesn't have to act according to righteousness and justice.

However, we've already seen that this model admits we can't understand this definition of grace because it doesn't make sense. That's why we would need the "golden key" Faith Model.

Grace Examples

> "Let no corrupt communication proceed out of your mouth, but that which is good to the use of edifying, that it may minister grace unto the hearers" (Ephesians 4:29).

> "I do not frustrate the grace of God: for if righteousness come by the law, then Christ is dead in vain" (Galatians 2:21).

We can give grace to others and we can frustrate the grace of God. If our model of grace comes from our non-contradictory definitions, these passages read differently.

The first passage is saying that our words to others that result in a benefit are grace to the hearers. We are the influence and the hearers make the choice as to whether they will let it influence their hearts and come out in their actions. It is a beautiful example of what God's grace is to the individual and an opportunity for us to be more like God (Figure 1).

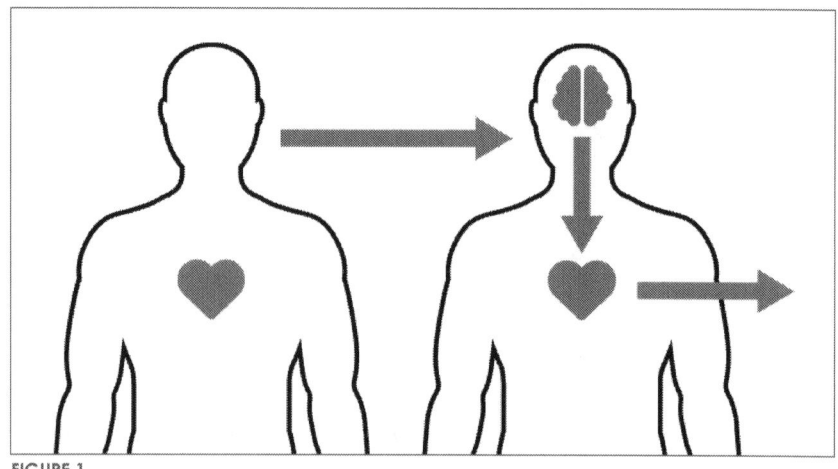

FIGURE 1

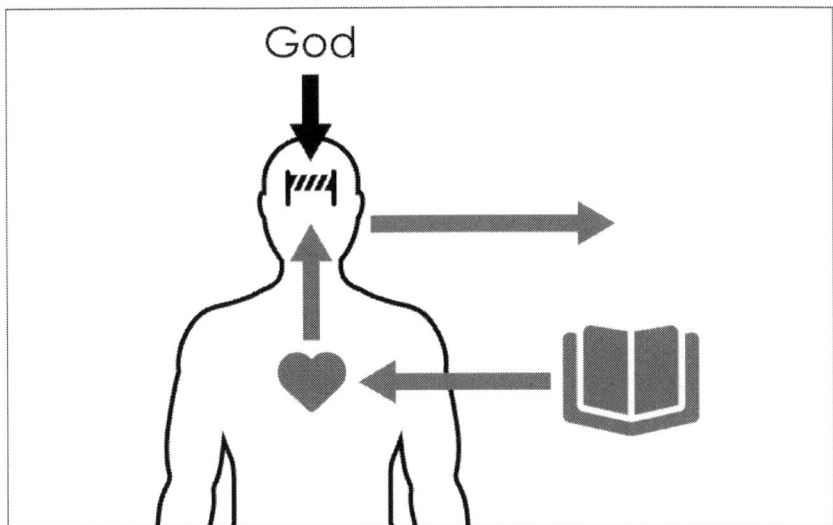
FIGURE 2

The second passage says that God's divine influence to the individual can be frustrated if the individual is looking to the law for righteousness (Figure 2). In fact, it says that Christ died so that God can influence us individually in order to make us righteous. Remember, grace occurs between God and the individual. The law treated everyone the same. Notice, in Figure 2, everyone gets exactly the same information (law), and "righteousness" comes from the individual.

Understanding and Grace

As for us being able to understand grace, if we weren't supposed to understand it, then there would be no reason for people to try to explain it to us in the Bible. Here are more people trying to help us understand grace, with emphasis added:

> "*Grace* and peace be multiplied unto you through the *knowledge* of God, and of Jesus, our Lord, According to his divine power hath given unto us all things that pertain unto life and

godliness, through the *knowledge* of him that hath called us to glory and virtue" (2 Peter 1:2, 3).

"Ye therefore, beloved, seeing ye *know* these things before, beware lest ye also, being led away with the error of the wicked, fall from your own steadfastness. But *grow in grace*, and in the *knowledge* of our Lord and Saviour Jesus Christ" (2 Peter 3:17, 18).

"For the grace of God that bringeth salvation hath appeared to all men, *teaching us* that, denying ungodliness and worldly lusts, we should live soberly, righteously, and godly, in this present world" (Titus 2:11, 12).

These verses say we are supposed to grow in grace and understanding by acquiring knowledge. We can see the "golden key" model of grace contradicts these Bible passages, while our non-contradictory model of grace as being something we have the power to frustrate, give to others, and intentionally grow in, fits perfectly. The last passage says that grace teaches us. How would "unmerited favor" teach us?

Let's look at some passages that more specifically show grace applies to actions for specific situations.

Action Grace

"Let your speech be always with grace, seasoned with salt, that ye may know how ye ought to answer every man" (Colossians 4:6).

If this were unmerited favor, our speech would be the same positive message to everyone about favor they don't deserve; however, "seasoned with salt" means the response may not be completely positive and needs to have the bitterness removed. In addition, the only way to know how "ye ought to answer" is if you are letting your heart be divinely influenced from God and letting it come out in your speech (actions).

> "Let us therefore come boldly unto the throne of grace, that we may obtain mercy, and find grace to help in time of need" (Hebrews 4:16).

This makes grace out to be something you can intentionally go looking for. Besides, how would unmerited favor specifically help a person in time of need? This passage says grace must be some specific understanding God gives us about what we should do in a specific time of need.

When people say grace is "assigned advantage," it implies that God gives some more grace than others and there's nothing we can do about it. God is not a respecter of persons (Acts 10:34). Our non-contradictory model says grace is available to anyone who wants it. It is not withheld from some more than others. Some have more grace and some fail grace, but it is because of the individual, not because of God, otherwise, it wouldn't be right or just.

Failing Grace

> "Looking diligently lest any man fail of the grace of God; lest any root of bitterness springing up trouble you, and thereby many be defiled; Lest there be any fornicator, or profane person, as Esau, who for one morsel of meat sold his birthright. For ye know how that afterward, when he would have inherited the blessing, he was rejected for he found no place of repentance, though he sought it carefully with tears" (Hebrews 12:15–17).

This passage shows that we can fail the grace of God. Esau failed grace because he wasn't diligent and made a bad decision. Notice there was nothing that could undo it. If grace were infinite unmerited favor from God, and God's responsibility, Esau wouldn't be at fault and the situation would be fixable.

> "Do ye think that the scripture saith in vain, The spirit that dwelleth in us lusteth to envy? But he giveth more grace. Wherefore he saith, God resisteth the proud, but giveth grace unto the humble. Submit yourselves therefore to God. Resist the devil, and he will flee from you" (James 4:5–7).

The "more" in this passage refers to the fact that there is more grace available to us than the evil in us. Notice what prevents grace. Grace resists the proud who won't put God first and listen to Him. It is given to the humble because they put God first. It takes humility to willingly forgo your own desires and allow God to direct your actions. No wonder the proud don't have grace. They are unwilling to listen to God and do it His way. They are unwilling to be obedient to God. Grace can't overcome the sin of pride.

The "golden key" definition of grace puts the responsibility on God. The New Testament definition of grace puts the responsibility on the individual. This is where the confusion with works occurs.

Works

> "But wilt thou know, O vain man, that faith without works is dead? Was not Abraham our father justified by works, when he had offered Isaac his son upon the altar? Seest thou how faith wrought with his works, and by works was faith made perfect?

And the scripture was fulfilled which saith, Abraham believed God, and it was imputed unto him for righteousness: and he was called the Friend of God. Ye see then how that by works a man is justified, and not by faith only. Likewise also was not Rahab the harlot justified by works, when she had received the messengers, and had sent them out another way? For as the body without the spirit is dead, so faith without works is dead also" (James 2:20–26).

"Even so then at this present time also there is a remnant according to the election of grace. And if by grace, then is it no more of works: otherwise grace is no more grace. But if it be of works, then is it no more grace: otherwise work is no more work" (Romans 11:5, 6).

In the first passage, James is saying works justify people. In the second passage, Paul is saying if works save us, then it is not by grace. How do we solve this apparent contradiction? It all comes back to understanding cause and effect.

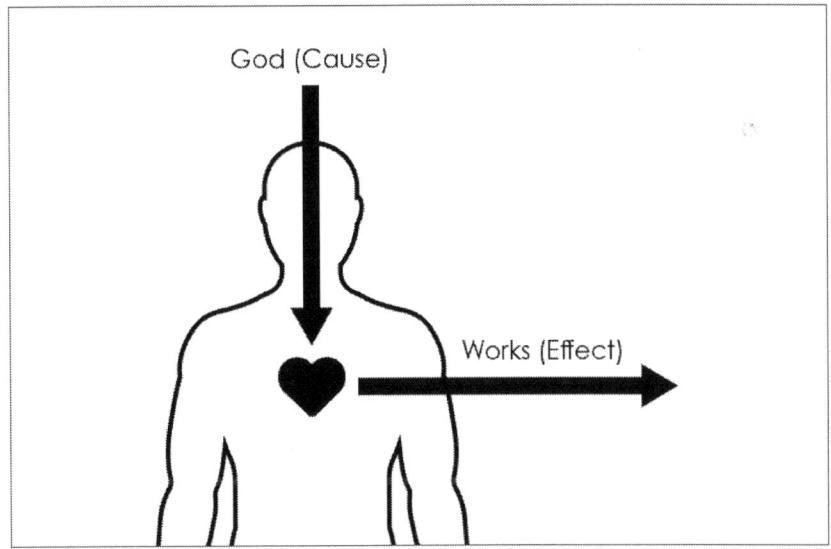

FIGURE 3

We are saved by causes (i.e., grace and faith). Works are the specific results (effects) of grace and faith. The actions (works) don't save us—the causes do (Figure 3). We will see works in people who have the causes. Faith and grace are made perfect when they lead to works; otherwise, it isn't faith or grace. This is the point of the first passage.

Take a moment to allow this point to sink in. Works prove and justify faith and grace. If a person believes they are operating in biblical faith and grace, yet they lack the actions that naturally result from these causes of salvation, they don't really have grace and faith—no matter how much they try to convince themselves and others.

Remember though that there may be more than one cause that results in an observed effect. In this case, works aren't solely a result of grace and faith. If someone has the effects, they could have them for the wrong causes (Figure 4). They may even decide the effects (works) are what save them. This is the point of the second passage.

For example, people who decide to be overly polite and treat everyone and every situation the same are relying on works. They think the

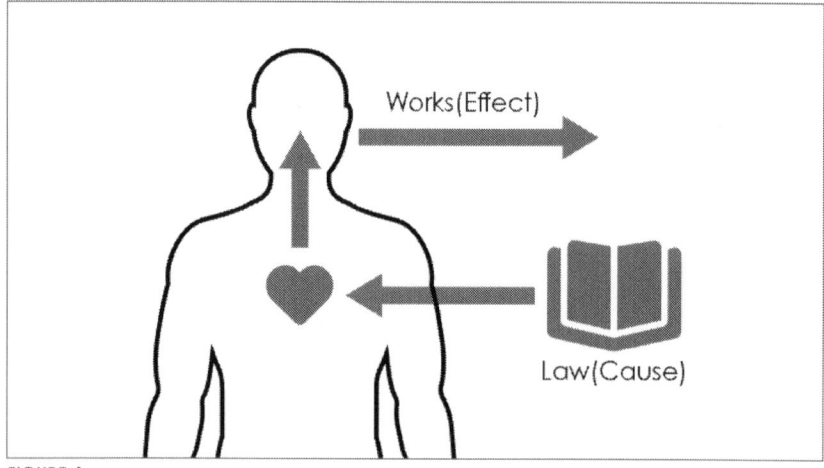

FIGURE 4

action is righteous in and of itself. Rather than justifying grace, they are frustrating it. We saw this with people who looked to the law as a cause (Figure 2). This is the opposite of grace. If they were following Biblical grace, they'd let God tell them how to interact with each individual person and situation. In some cases, God may tell the individual to be less than polite and to season their response with salt.

Wisdom

James also has something interesting to say about wisdom, writing, "If any of you lack wisdom, let him ask of God, that giveth to all men liberally, and upbraideth not; and it shall be given him" (James 1:5).

James is saying you can jump over knowledge and understanding in order to get as much wisdom as you want. Is this just? How is this possible?

We saw that wisdom is the ability to make good decisions. This passage says you can ask God how to make a good decision and He will show you. Isn't this the definition of charis? This passage is another example of how we can go to God at any time and get as much direction as we need in order to make a good decision that is proven in our actions because it is grace.

Summary

Paul is saying that people who do the right things (effects) for the wrong reasons (causes) are counting on works (effects) to save them (Figure 4). Paul spends a lot of time explaining this. The reason these passages are confusing is that people take the "golden key" model of grace and think a specific effect can only come from one cause.

With the "golden key" model, grace saves us regardless of what we do, regardless of our works. In fact, "golden key" faith and grace do not account for actions in their definitions. "Golden key" grace begins with you taking a wrong action and ends with God giving you value. Biblical grace begins with God giving you a value (gift) and is made perfect with you taking a right action by letting God work through you.

Paul is saying that works are valuable, but they don't save you. The only way works can be important but not the reason for your salvation is if works are an effect of the cause that saves us—the same conclusion resulting from our non-contradictory definitions and models of faith and grace!

- Reread all the verses in this chapter and try to make "unmerited favor" or any other definition fit. Can you do this without contradiction?
- How does this definition contradict the idea of "once saved, always saved"?
- What are the implications of God speaking to us individually?
- What works (effects) prove you are operating in grace (cause)?
- What is your Salvation Model?

Joel Swokowski's *Commentary*

The author went through an arduous process in the writing and editing of this book, even meeting with seasoned pastors and theologians to get their take on what was being written. Lenhart had a discussion with an experienced pastor about this doctrine of grace, and that pastor even agreed with the definition of grace presented in this chapter. Yet the pastor stated that he would continue to preach grace as "unmerited favor," rather than having to tell his congregation that he was wrong. Really, this was an effort to maintain an appearance of being right and an appearance of not being wrong. Something I've experienced in my years interacting with pastors like this is they tend to confuse "being right" with "righteousness." Unfortunately, holding onto their appearance of being right meant they would rather people miss out on salvation than admit they were wrong.

Lenhart was also accused of not giving Jesus enough credit in this book although there's a statement in this chapter that states, "In fact, it says that Christ died so that God can influence us individually in order to make us righteous." For me to accuse Lenhart of this, I would have to either blatantly ignore parts of this book, or I would be proving my entire lack of ability to interpret a text, including the Bible!

Thorn in the Flesh

In 2 Corinthians 12:6-9, Paul shares the following,

> "For though I would desire to glory, I shall not be a fool; for I will say the truth: but now I forbear, lest any man should think of me above that which he seeth me to be, or that he heareth of me. And lest I should be exalted above measure through the abundance of the revelations, there was given to me a thorn in the flesh, the messenger of Satan to buffet me, lest I should be exalted above measure. For this thing I besought the Lord thrice, that it might depart from me. And he said unto me, My grace is sufficient for thee: for my strength is made perfect in weakness. Most gladly therefore will I rather glory in my infirmities, that the power of Christ may rest upon me."

We don't know what Paul's "thorn in the flesh" was, and it is often discussed and debated. Yet, if you look clearly at the passages above and the rest of 2 Corinthians 12, you'll see that Paul was not given this thorn because of anything he did wrong or any sin he committed. Furthermore, the Lord's answer to Paul on how to deal with the thorn was His grace. This once again shows us that grace cannot be defined as unmerited favor. Paul was not doing any behavior that could be construed as "unmerited." The rest of this story is covered in chapter 15.

Further Commentary Regarding the Reference to James 4:5-7

James referenced the scripture to show that God resisted the proud. The only way God can grow a person not flowing in grace is from the outside, from resisting them. However, God gives more grace to the humble. God grows the humble person more from the inside. Believers are humble and will be able to speak more and more for God through grace. God has provided everything necessary for salvation. The proud people are at fault when they do not receive grace. Pride is "unwilling to consider you are wrong and in need of nothing." Someone in pride is not going to reflect God's influence in their life, they are going to present their own influence which is actually from the world.

Humility is "the ability to consider a perspective other than your own, including that you could be wrong." This ability is contrastive thinking. Just as this chapter states, "it takes humility to willingly forgo your own desires and allow God to direct your actions." It takes humility to receive the influence that God is sending your way.

I covered James 2:14-26 at the end of Chapter 6 to look at the relationship between faith and works. This chapter showed more clearly why the definition of Causality needs to be defined in one direction. Faith is a cause that will always result in works. However, it is possible for a person to have works without faith being the cause.

CHAPTER 10

Uniqueness

BEFORE WE FINISH uncovering the Salvation Model, we need to understand an important implication of our non-contradictory definitions: uniqueness. Remember, biblical grace states God is influencing the hearts of individuals.

We are not under the law that treats everyone the same. Grace doesn't give us all one set of rules to live by. The non-contradictory definition of grace says that a person needs to let God carry out the action He is telling them to do for their situation. Therefore, our actions depend on the situation and our uniqueness.

One consequence is that this action may not be what God is telling others to do in the same situation. In fact, God might be telling others to do something different than what He is telling you. God needs to treat everyone differently because each person is unique.

"Uniqueness" is the belief we are all made differently for a different purpose or use.

> "Hath not the potter power over the clay, of the same lump to make one vessel unto honour, and another unto dishonour?" (Romans 9:21).

Dishonor doesn't mean useless. For example, some pottery is for display purposes, while other pottery is for utilization. It is up to the potter whether they want the lump of clay to serve a delicate or hearty purpose. Like pottery, some people were made to carry out delicate functions, while others do the less glamorous work. This aspect of uniqueness is qualitative.

Qualitative

When I discuss the God Model with people, one of the first objections I get is that God can't be defined because He is different to different people. I like to explain this by seeing us as ships on the ocean with God as the lighthouse.

Every ship sees the lighthouse. Every ship communicates with every other ship the location of the lighthouse relative to their location. However, no two headings are the same; one sees the lighthouse due east while another sees it due north. At this point, it looks as if there are two possibilities: 1) There is only one lighthouse, or 2) There is more than one lighthouse.

If the ships all originated from the same point, then there must be multiple unique lighthouses. If the ships originated from different points, then it is possible there is one lighthouse. When people say God can't be defined, they are denying uniqueness. They are really saying people are the same, so there must be many God Models.

If people focus on justifying themselves, then they are going to conclude there must be more than one lighthouse. If they focus on justifying God, then they are going to conclude there must be one God and the different perspectives are a result of the uniqueness of people.

The God Model presented in Part I is the result of discussing God with people from many walks of life for four years. When I focused on justifying God, I realized there must be higher principles that can account for the different perspectives people have of God. I like to think that God made it so that we would have to discuss Him with others in order to understand Him more.

Paul likens this qualitative uniqueness to a body where each part is necessary and serves a different purpose. Likewise, we can't all do the same things, but we are all necessary.

> "For the body is not one member, but many. If the foot shall say, Because I am not the hand, I am not of the body; is it therefore not of the body? And if the ear shall say, Because I am not the eye, I am not of the body; is it therefore not of the body? If the whole body were an eye, where were the hearing? If the whole were hearing, where were the smelling?" (1 Corinthians 12:14–17).

Remember, the law saw everyone as the same. Paul is saying being seen as different is not a bad thing; in fact, it results in more value for everyone. Everyone benefits from the abilities of others. Paul continues, "But now hath God set the members every one of them in the body, as it hath pleased him. And if they were all one member, where were the body? But now are they many members, yet but one body. And the eye cannot say unto the hand, I have no need of thee: nor again the head to the feet, I have no need of you" (1 Corinthians 12:18–21).

Again, Paul shows that God has a different purpose for each person depending on the unique gift he's been given. God set the members in the body. God is "the potter" in the first passage. This passage shows that God sees us as unique beings; however, not only do we have different talents, but we also have different quantities of each talent. Let's look at the quantitative aspect of uniqueness.

Quantitative

> "For the kingdom of heaven is as a man traveling into a far country, who called his own servants, and delivered unto them his goods. And unto one he gave five talents, to another two, and to another one; to every man according to his several ability; and straightway took his journey" (Matthew 25:14, 15).

God gave (gifted) each individual a different number of talents from "his goods." Again, we see God is not only determining the uniqueness of the individual, He is also the source.

> "Then he that had received the five talents went and traded with the same, and made them other five talents. And likewise he that had received two, he also gained other two. But he that had received one went and digged in the earth, and hid his lord's money" (Matthew 25:16–18).

Two of these individuals used the talents they were given to produce more. One individual hid his talent.

After a long time, God returns to find out what they've done with their talents. God compliments the two individuals who have used their talents to produce more. God says their actions show they are faithful and He says He will make them rulers. The individual that hid the money is a different story. The Bible says, "Then he which had received the one talent came and said, Lord, I knew thee that thou art an hard man, reaping where thou hast not sown, and gathering where thou hast not strawed: And I was afraid, and went and hid thy talent in the earth: lo, there thou hast that is thine. His lord answered and said unto him, Thou wicked and slothful servant, thou knewest that I reap where I sowed not, and gather where I have not strawed: Thou oughtest therefore to have put my money to the exchangers, and then at my coming I should have received mine own with usury" (Matthew 25:24–27).

The individual who did not use his talent tells God his excuse for not producing anything. God used justice by showing that the individual's excuse should have led him to a different action. In fact, God even points out the servant knew something, but his actions didn't follow what he thought he knew. God is still focused on thinking (understanding) and the individual's thought process.

God calls the servant wicked and slothful. Then God finishes His discussion with the servant by telling him what he ought to have done in order to be righteous. Now God is able to execute judgment, and says, "Take therefore the talent from him, and give it unto him which hath ten talents. For unto every one that hath shall be given, and he shall have abundance: but from him that hath not shall be taken away even that which he hath. And cast ye the unprofitable servant into outer darkness: there shall be weeping and gnashing of teeth" (Matthew 25:28–30).

At the end of the story, God sends the individual to hell. Notice, this had nothing to do with the amount of value God gave to each. It had everything to do with what the individual did with the value given to them. The individual who gained two talents did not do as well as the individual who gained five talents, but they both made it into the joy of the Lord.

In fact, God told the individual with the one talent that he would have been fine if he had just made interest on the talent. Instead, the individual was punished because he did not gain anything with the talent, although he had the ability to produce more with the talent. Actually, the servant returned less to the master in much the same way as a quarter from a long time ago was more valuable back then than a present-day quarter is today.

Expectations

We all know you can't have the same expectations for different people—for instance, you wouldn't expect me to play golf as well as a professional golfer. The real damage, however, is done when we put these expectations on ourselves.

When I play a round of golf with the expectation that I will shoot par, I end up shooting at least thirty shots worse than par. Every time I have only one or two shots in order to make par on a hole, I end up attempting something beyond my abilities. The result is several strokes greater than par. When I tell myself I have an extra shot on every hole—my objective becomes eighteen shots more than par—I end up playing within my ability and finishing the round with a score that is only ten shots over par. Unrealistic expectations adversely affect our results.

While God can do miracles through us, we need to be realistic about our expectations of ourselves. God does not have expectations of you greater than what you can do. God is realistic about His expectations for you because of your uniqueness. Additionally, God does not have the same expectations for everyone.

Jesus and Paul are saying there are different expectations on different people according to abilities given and purpose for their life. Isn't this consistent with justice? Why would it be just to expect some people who have been given less to have to achieve the same (either qualitatively or quantitatively) as others who have been given more in order to be valued by God? Clearly, it is just for the value God expects from us to depend on the amount and abilities God has gifted us.

Likewise, salvation is dependent on our will through making an intentional effort to grow in grace and faith. There is nothing in the parable of the talents story that glorifies *not* thinking. There is nothing in the story that says the individuals made it to heaven due to unmerited favor. The individual must choose to accept the purpose they were made for and actively use this talent (gift) to yield a profit.

See how this is different from "golden key" grace, which says God gives everyone different value that they didn't deserve to cover sins for salvation. This grace completely depends on God and is out of the control of the individual. Again, this grace begins with a bad action by the individual and ends with God giving a value to the individual. Biblical grace begins with God gifting a value to the individual and is made perfect when the individual chooses to use this value to create more value (righteousness), exactly like in the parable we just examined.

Jesus and Paul are saying that God has given everyone different abilities and they are judged on their ability to return a just and righteous value. God doesn't give anyone an assigned advantage over anyone else. The value (talent) is on an individual basis and is not affected by the value given to others. The value is not confined to salvation and covering sin. The value is dependent on the individual's ability and purpose.

Summary

Things that are the same are separate; things that are different can be connected. Examples are two nails and parts of the body. The word "same" implies there is more than one of something, such as nails. Because there is more than one, they are separate by definition. If they were joined, then they wouldn't be the same. Parts of the body can be joined together because they are different. So, it is our uniqueness that allows us to connect to others.

People deny uniqueness when they say everyone has to do the same actions. This is not right, nor just. They usually abuse cause and effect to do this. They want everyone to have the same cause, but they can't control the inside of the individual, so they focus on trying to get the same effect out of everyone by making everyone do the same works. This is known as "legalism." They are denying the importance of the attitude in the individual's heart, or more to the point, the reason (cause) for the effect. They are denying biblical grace.

Some people are concerned that believing the God Model and Salvation Model will hinder their uniqueness. Actually, these models result in the individual becoming more unique and more the person God created them to be, because it is God that flows through them. This book is really foundational. Allow me to explain.

When people drive down a street, they notice the differences between the buildings. Each building is unique. People don't look at the foundations. In fact, the only time anyone notices a foundation is when it is faulty. Likewise, the principles in this book are meant to help you fix your foundation so that God can build something unique on top of it.

These principles also quickly expose the flaws in other people's foundations. Furthermore, you may notice that a significant number of speakers spend their time disguising their faulty foundation rather than instructing you on how to improve your foundation. Unfortunately, there are speakers who focus on making everyone's building the same.

This denies uniqueness. Jesus didn't deny uniqueness. He related to people differently.

Now that we understand faith, grace, and uniqueness, we can identify the Salvation Model.

- What is unique about you?
- What could be some reasons God created you in that way?
- Thinking about yourself and another family member, why is it good you are not the same?
- Can you describe how each unique attribute is a benefit to the family?
- How does the law deny uniqueness?

Joel Swokowski's Commentary

This is a crucial chapter that can serve as a rest stop on our way to understanding salvation.

We are not under the law where everyone was treated the same. We are under grace, where God speaks His influence to the individual. Let's look at an example from Hebrews 8:10-11.

Verse 10 says, "For this is the covenant that I will make with the house of Israel after those days, saith the Lord; I will put my laws into their mind, and write them in their hearts: and I will be to them a God, and they shall be to me a people:"

The Greek word for "mind" in this verse was "dianoia." It was composed of two Greek words that meant "the channel through which the Mind/Soul operates," which is the conscious brain! The word brain is a relatively new word and was created more than a thousand years after the New Testament was written. The laws being written on the heart would be written as an effect of the person intentionally choosing to allow this influence in the conscious brain to come out in their actions, which is grace!

Verse 11 continues, "And they shall not teach every man his neighbour, and every man his brother, saying, Know the Lord: for all shall know me, from the least to the greatest."

God's guidance would not be outside of the individual through the Law. It would be inside the individual in their conscious brain and upon their heart as an effect of them choosing to allow these actions to come out. When I see the fact that God's influence first enters into a person's brain, it's even more clear that God's influence is going to be different for everyone. My brain is as different from yours as it is from everyone elses' who has lived.

The lighthouse analogy explains even further how people hear from God differently. I've heard time and time again that "God's too big to be put in a box," yet these same people seem to believe that God is very limited in how He speaks to us. What would you say if I told you God speaks to me through a tree? Most people would think I was not hearing from God. Does it have to be a burning bush? The Bible showed ways God spoke to people that most traditional pastors would call heretical.

If a pastor did this to a person who was hearing from God, then they would be actively hindering the person from hearing from God. They would be encouraging the person to not act in grace!

Clearly, this person would fit the definition of a stumblingblock, the same way it would have been a stumblingblock to tell Moses he wasn't hearing from God when God spoke to him through a bush! Would you have wanted to be the person to tell Moses that he wasn't hearing from God? Would you have wanted to be the person who diminished Moses' faith?

Although God speaks differently to everyone, one thing we can all rely on is that He will remain Right and Just in all the influence He gives. Since we have an objective measure for God, we would be able to help the person determine if they were hearing from God or not.

The difference is not in God being God (Right and Just) towards people. The difference is in how each person is unique in who they are, their thought process, and the path they are walking.

This chapter brought the faith and works discussion to a dramatic conclusion with an application. Not understanding the causal relationship between faith and works results in people denying uniqueness. They think the works are the cause and they end up damaging the person by taking them out of their uniqueness.

CHAPTER 11

Salvation Process

THE "GOLDEN KEY" definitions looked at salvation as a one-time event. You pray "the prayer" one time and you are in for good, just like a person giving you the key to the party. However, we know you haven't been saved yet. You have a will and could still renounce everything.

Salvation occurs once the individual dies and makes it into heaven. Once you make it to heaven, you don't have to worry about getting kicked out because "once saved, always saved."

The non-contradictory definitions of faith and grace define a process. Since we are saved by grace through faith, salvation is also a process. Faith and grace are actions you choose to do intentionally and in which you can continually grow. Likewise, salvation is something done intentionally and in a manner in which you continually grow. So, the process of salvation is similar to the process of becoming more righteous and just.

Now we completely understand the key passage that says, "For by grace ye are saved through faith" (Ephesians 2:8). "Saved" refers to the result at the end of the process of salvation. If "saved" referred to a current result, then it would contradict either your will or the concept of "once saved, always saved."

Let's look more closely at the salvation process by addressing some common questions.

How Did God Justly Obtain Infinite Value?

We've already seen how God used justice to obtain the infinite value needed to cover our sins. Jesus lived a sinless life and would have lived forever. He lost this ultimate value, His life, without just cause. Justice required Jesus to receive infinite value.

This value God needed to obtain had to be infinite. God could not create just enough value to pay for everyone because God doesn't know how much each of us needs until we've made choices. Remember, we have free will. We determine how much or little we are going to deviate from God's holy plan for our lives. God knows how much we need after we make our choices, but because God is outside of time, He sees every moment at once.

God doesn't know which people are going to need the value until after they have expressed their will to focus on the spiritual over the physical. Therefore, God could not justly identify that exact amount ahead of time. Also, God had to generate enough value to cover everyone who has lived or will ever live. God had to generate infinite value.

Remember, Jesus is God in that He is the embodiment of what is righteous and just. However, Jesus was also a man in that He had a free will. He could have chosen to do anything. Instead, He chose to do everything according to righteousness and justice. That is what makes Him perfect. Once Jesus obtained this value, He freely gave it to God the Creator.

How Does God Give Us the Value?

Jesus gave that value to God, and God has given everyone access to this value. It is a gift. We have to choose to take this value, however—we have to express our will. It would be unjust of God to credit this value to people who don't want it—that would be God violating their will. Remember, the "golden key" definition of grace said there was unlimited value given to people God chose and the people didn't have to do anything to get it.

Withholding this value from people who wanted it would be unholy. Giving them a value without requiring a value in return would violate justice. So what must we do to receive this gift? This value is obtained through the process of confession and repentance.

The only reason we need the value is due to sin. Confession is the admission of what we specifically did that was wrong. Until we admit what we did, we will have guilt. We will be defensive, waiting for someone to point it out. Admitting guilt leads to freedom.

Repentance is the turning away from the action. This proves the will of the individual. Do they really want to stop doing it and put God first or do they want to keep doing it? Again, until we make the effort to turn away from sin, we will experience guilt.

Notice it takes words (confession) and actions (repentance) in order to get the value, just like faith and grace were shown through word and actions, and just like the salvation process is initiated through confessing with your mouth (word) and believing in your heart (action). Also, confession and repentance are intentional actions that are in our control.

Finally, notice that confession and repentance mesh with the process of becoming more right. Remember, in order for you to become more right, you must admit you are wrong (confession) and change your beliefs (repentance). God gives us value because we are going against our nature and trying to become more like God (righteous). Recall, righteousness is the requirement for salvation. Confession and repentance are righteous and remove unrighteousness.

What is Sin?

Paul defines sin as, "[F]or whatsoever is not of faith is sin" (Romans 14:23). Now we can see why it is so important to have a non-contradictory definition of faith. Faith believes in something we can't see or that hasn't happened yet. Faith looks long term because it knows the causes. It doesn't focus on the effects and gives in to a short-term mentality. Faith is built through understanding and experience.

Therefore, sin is anything we do that does not take into account the long-term (eternal) aspects. Sin is an action, attitude, or thought that is wrong because it is done apart from understanding and experience. That is to say, anything done apart from what the individual knows is right due to their understanding or experience.

> "Therefore to him that knoweth to do good, and doeth it not, to him it is sin" (James 4:17).

This is not an objective list of actions, like the law. It is dependent on the individual. If it is wrong for the individual, then it is sin for the individual alone. This is consistent with uniqueness. There are other definitions of sin that deny uniqueness and lead to condemnation.

For instance, some people define sin as "missing the mark" or "anything that doesn't bring glory to God." These definitions imply that everything is either a sin or something that is perfect and brings glory to God.

It's like flipping a coin and then wondering if my coin flip brought glory to God or if it was a sin. Having this type of definition for sin leads people to feel condemnation when they spend even one minute doing something that isn't bringing glory to God. These definitions are simply tools for people to focus on others. Paul's definition leaves it to the individual to focus on himself.

Besides, it is not our responsibility to point out what is sin in other people. If we judge others unjustly, justice will require a value from us. We shouldn't be concerned with whether others are "getting away with something." Sin causes guilt. If it is sin for the individual, the individual will have guilt. People who sin are getting a penalty whether you see it or not.

Actually, it is holy of God for us to be designed so that sin causes guilt. Our goal should be to follow God's influence in our life. When we don't operate in grace, we should desire to know this. Guilt is our objective way of knowing we aren't following God.

The individual's response to guilt becomes an expression of their will. Do they want to follow God, understand more, and remove the guilt through confession and repentance? Or do they want to follow their flesh, not think, and try to ignore or transfer the guilt?

What Benefit is Confession and Repentance Before We Die?

Besides helping us to become righteous and facilitating the salvation process, confession and repentance also leads to an opportunity to right our wrongs. Earlier we saw that suffering can be an effect of something we cause. We can limit our suffering and the suffering of others by confession and repentance.

Confessing to people we've wronged benefits all parties. First, it benefits the person who was wronged by identifying the cause of his or her suffering. Secondly, it allows both parties to focus on taking action to right the wrong—that is, repentance.

If we are truly repentant, we will ask the person we have wronged what we should do to pay for the wrong. Notice, if the wronged party asks for too much value, we should pay and rely on justice to make it equal.

For example, my daughter came to me one day upset that her brother licked three of her lollipops. My son was six at the time, but my wife and I had been teaching these principles to him since he was born. When his sister and I confronted him, my son confessed and repented. He said, "I licked the lollipops. How can I make it up to you?"

At this point, his sister made a request that caused tears to well up in his eyes. My little man looked at me glassy-eyed as if to say, "This hurts to do the right thing."

I put my arm on his shoulder, looked him square in the eye, and asked him, "What happens if her request is greater than what you think you should have to pay?"

Slowly, a smile came over his face as he said, "Then she'll end up owing me because of justice."

Instantly, she grabbed him and blurted out, "Wait, let me ask for something different!" She realized justice would cause her to end up owing if her request was too onerous.

Finally, the wronged person has the option of forgiving us and taking no payment. Forgiveness is the expression of the will by the wronged person that he is not going to try to equal out justice himself even if he has the opportunity to equal out justice. This is focused on action and not emotion.

Notice, forgiveness doesn't require the wronged person to feel good about the offender or forget about the offense. However, both of these things will result in greater assurance that the offended person won't have a change of heart and try to equal out justice in the future. If they truly forgive us, then God will credit them with the value in order to equal out justice. Even if the wronged person takes a value from us, they need to forgive us, or justice will eventually require a value from them (Matthew 6:15; Mark 11:26).

Forgiveness and justice are two separate issues. Forgiveness allows justice to occur. When we ask people to forgive us, we are really asking for them not to equal out justice so that God can do it. In the long term, forgiveness benefits the forgiver, not the forgiven. Rather than focusing on getting forgiveness, the offender should actually focus on actions that give the offended person value in order to equal out justice.

Likewise, this process is the same for our interactions with God. Now we see why God is quick to forgive us even before we confess and repent. If God didn't, He wouldn't be righteous and just. If God didn't forgive us, then justice couldn't act. God's forgiveness doesn't violate or prevent justice; it actually causes justice to occur. This fact proves that God forgives us instantly, but that we don't receive the benefit and equal out justice until we confess and repent. God does this with every sin except blasphemy of the Holy Spirit (Matthew 12:32; Mark 3:29; Luke 12:10).

Why Does Confession and Repentance Solve Our Problems?

First of all, the problem is guilt. When we act apart from holiness, we experience guilt. Unholiness is the cause and guilt is the effect. God has built everyone in a manner that causes us to experience guilt when we've sinned. This is done so we know we need to take action. How we respond to guilt is in our control.

As humans, we have a need to remove this guilt, and this leads to all types of strange behavior. Some people convince themselves they can earn the right to remove guilt by punishing themselves. Whether this is reciting words or committing physical acts, this cult-like behavior is meant to relieve the individual of guilt.

Some people respond to guilt by trying to blame others. In fact, it can lead these people to do harm to others who are innocent in hopes of convincing themselves that it is someone else's fault. Still others will fill their lives with activities so they don't have time or are unable to think about their guilt. For example, there is nothing wrong with hobbies or alcohol in and of themselves. However, when these activities are used to prevent the individual from dealing with his or her guilt, these activities can be the cause of failing grace.

As you can see, guilt becomes a cause. Ultimately, guilt results in pain, and pain results in fear. That explains the reason behind the psychiatrist's process of helping the individual. They start with the fear and work backwards until they identify the guilt. It is the guilt that is the source of all human problems.

I see guilt as an open sore on the person's spirit, with pain oozing out of this open sore. People need to displace this pain. We see this when someone suddenly "goes off" on an unsuspecting person. What they are doing is displacing their pain. Unfortunately, the unsuspecting person now has pain, and if they don't handle it well, they will end up displacing it on yet another unsuspecting person.

A healthy example of displacing pain is when we talk with our friends. Things that are tremendously painful to us can be vented to a listening friend. We feel better if the friend has agreed to bear this pain. People who bear pain do a tremendous service. If they didn't, the guilty people would experience fear and their actions could end up permanently injuring innocent people. Ultimately, the best way to get rid of pain is to let God bear it.

Even if we help people get rid of pain, we haven't dealt with the problem. We have only dealt with the symptom. The problem is guilt. How do we help people heal this open sore? The only way is through confession and repentance. Anyone can bear pain; only Christians know how to remove guilt because it requires God.

The ultimate cause of guilt is unholiness (sin). It seems holiness is the solution to all human problems. Grace is God's solution to our lack of righteousness. If we let God influence our decisions, we would never be wrong. We would be holy—except we make mistakes, so we will never be holy. Only God is holy.

The Bible says in Romans 3:23 that all have sinned. We will never be able to avoid sin; however, we can be free of guilt. This is where our focus should be placed. Remember, confession and repentance are the only ways to remove guilt.

The world suffers from sin because it doesn't operate in grace. This lack of holiness leads to guilt. The entire world suffers from the inability to remove guilt, and we have the answer. This is the first step to freedom. So far though, Christians have been terrible salespeople.

An effective salesperson understands his product, believes in his product and understands the customer's needs. While Christians certainly appear to have enough passion to believe in their product, they don't understand what their product is and what the "customer" really needs. Once we have a non-contradictory model, the answer is simple.

The problem is guilt. The "customer" needs a way to live without guilt. Holiness prevents guilt. While grace is the ultimate answer and the second step to freedom, people are still going to make mistakes and sin. If the individual has made decisions apart from holiness, they can use confession and repentance to remove the resulting guilt. This sounds simple. Why don't the people of the world confess and repent?

How Does the World Try to Remove Guilt?

Man has a way of circumventing this process, and it is called "apology." An apology is when people say they are sorry or feel bad. Most apologies don't consist of people admitting what they did. Most apologies don't involve promising they won't do it again.

Listen to apologies closely. Most sound like this: "I'm sorry your feelings were hurt," or, "I'm sorry this bothered you." Neither identified what I did. Neither said, "I'll never do it again." Some athletes have even given public apologies without stating what they were apologizing for or why! Apologies are focused on the short term. If anything, apologies are an attempt to receive forgiveness but ensure that justice will occur more quickly against the offender.

What value is an apology from the individual who has sinned against you? If the person hasn't admitted his wrong action or worked on not doing it again, there's no long-term value to you. Apologies don't help people become holier. Notice, there are no apologies in the Bible.

We have guilt when we do something we know is wrong. We need to remove that guilt. Apologies try to remove it, but they don't. Confession and repentance are the only ways to remove that guilt. That is what Jesus is offering—the ultimate way to remove our guilt.

What is the Party?

Looking back at the party analogy, we see that salvation is actually the same process as improving our relationship with God and intentionally getting closer to the party. So, the party is not salvation. Salvation is part of the journey to the party. For now, the party represents a relationship with God that is perfect. It appears to be something we will continue to work on once we are in heaven.

Notice that people who begin this salvation process and die shortly thereafter have not made it to the party, yet we know they are in heaven. Salvation must happen at any time along the journey when the individual dies, but the journey continues. It looks as if we are going to be on a journey that ends in this existence once we die, but must continue afterwards until we reach the party.

The journey will lead us to the party. Determining what specifically the party is will have to wait for the next book. In this book, we will look at some implications and finish uncovering the Salvation Model.

- Think of something you've always thought of as a sin.
- How would this action not be sin for someone else?
- What was the last incident that you apologized for?
- Did your apology include a confession?
- Did your apology result in repentance?
- What words and/or actions prove the confession and repentance?

Joel Swokowski's *Commentary*

This is a complicated and dense chapter because it is pulling together everything we have learned in order to present the model for Salvation. One thing to help clear this up is understanding what being "saved" means and what being "born-again" means.

Salvation, or being saved, is the avoidance of a punishment. We won't *achieve* this until we actually avoid the punishment, which doesn't happen until the day of judgment. Once a person is able to avoid that eternal punishment of God's Judgment, they are, at that point, saved.

Being "born-again" is the cause of achieving salvation. If a person is "born-again," their spirit is born and their soul is in a state of "salvation." If they continue living the "born-again" life, on judgment day, they will be saved. Becoming born-again, or beginning the Salvation Process, begins when a person confesses and repents. Think of the "born-again" experience... it's usually started with a prayer to God that sounds something like,

> "Dear Lord, thank You for sending Your Son Jesus to die for my sins. I've been wrong in how I've lived my life (confession) and I now submit to Your guidance (repentance) in hope of salvation. In Jesus' Name, Amen."

Notice, a prayer that leads to becoming born-again is really a confession and repentance to the Lord. It's an admission that I'm wrong, that I'm a sinner, and that I want to turn the other way by taking direction from God. I like to see this experience as the first time a person truly confesses and repents to the Lord.

Specifically, the person is confessing their plan for their life isn't working (I'm a sinner). Next, the person states they don't want to do their plan anymore (I admit that I don't have the answers). Finally, the person states they want God to do His plan through them via the Holy Spirit (I want to let Jesus in my heart). When the person makes these statements and means the sentences that preceded each one, they will experience transformation. If they state the sentences without meaning the rest, they may find themselves having to do this prayer over and over again. Why not just state specifically what you are confessing and repenting? Is it because people don't know specifically what they are confessing and repenting?

I can also see how this is the person living according to faith and grace, the causes of the salvation process. The confession would come due to my faith that I did something wrong (my plan), and the repentance would be me taking direction from God (grace) to make up for the wrong behavior.

Becoming born-again doesn't prevent us from the consequences of our actions. Justice and Causality still exist upon conversion. This was very true in my own life upon my conversion. I was unemployed, homeless, and living a life that I had never dreamed I would stoop to. When I received Christ as my Savior, I was still unemployed and homeless and dealing with the effects of the poor decisions I made before coming to the Lord. Yet now, I was pursuing the Lord's wisdom and taking direction from Him in an effort to make up for the injustices I committed. I was no longer isolated!

Credit vs. Responsibility

You will see throughout the rest of this book the difference between "credit" and "responsibility," including who gets credit and who is responsible as it relates to salvation. For now, know this:

1. God gets the *credit* for our salvation, He is the cause of this gift that He's given humanity.
2. Humans are *responsible* for their salvation. It's up to our choice to receive and apply this gift to our lives.

People tend to lump these two concepts, resulting in "experts" thinking there are only two possible explanations for salvation. If we get all the credit and have all the responsibility, then we are God. Clearly, this isn't a "Christian" explanation. If we get none of the credit and take none of the responsibility, then either everyone goes to heaven or God is unjust. This confusion between the concepts is where the two possible explanations that have caused division in Christianity occur.

Everyone going to heaven results in complete freedom for the individual. There are no requirements, and those evangelizing this religion tend to spend their time trying to explain the process for how people like Hitler could eventually make it to heaven.

The other explanation cannot support how a right and just God would create beings who were prevented from heaven and destined for eternal torture. Again, the discussion focuses completely on the explanation for this process because its existence is presupposed. The most destructive explanation is determinism: the future has already occurred and we are simply living out an existence that has been limited to only one possible day-to-day experience. Basically, God already knows what is going to happen and we not only don't have free will, we don't have freedom of choice. There are many passages that contradict this (God saying "Now I know…" when Abraham tries to sacrifice Isaac, Isaiah's prophecy about Jesus saying what He will do until He is able to **choose** the good and avoid the evil, etc.), however, the response is that we just can't understand it, which would make God unable to explain Himself to us.

How do we humans apply this gift of salvation to our lives? How do we receive this gift? Confession and Repentance!

Confession and Repentance

This chapter presented this very clearly, even bringing in the world's version: apologies. I cannot count the times I've heard a pastor advise a person who was wrong to "say you're sorry and ask for forgiveness." An apology is a man-made tradition that is an attempted loophole to God's standard for how to handle sin. How does me telling you how bad I feel help you? Isn't me asking for forgiveness just me trying to say, "please let me off the hook."?

If I confess and repent, I don't need to ask for forgiveness. The wrong would have been made right in my repentance. I can, and maybe should tell you how sorry I am or how bad I feel for having wronged you. However, the real benefit to you comes when I state "I was wrong. This is why I did the wrong thing. I don't want to do it again." I call this a "Full Confession" when all three statements are covered:

1. What I did wrong.
2. Why I did the wrong.
3. I don't want to do it again.

The confession ought to stop the sinning. The repentance is not just stopping the sinning, it is making progress in the opposite direction, beginning to make up for the injustice. Hopefully, to the point that the other person can look back on my sin with a good emotion. The proof that I truly repented is that I'm willing to talk about the incident long after it happened.

Sin

Another topic that the author was persecuted in, even called a heretic for, was sin. He was accused of teaching that sin is not an objective standard. Allow me to set the record straight: Lenhart did not teach this. Again, this was a gross misapplication and misinterpretation of this book. Sin is an objective standard, to the individual, due to the faith of the individual. Another way to see sin is "any *what*, with a wrong *how/why*."

We've seen Romans 14:23b says, "... for whatsoever is not of faith is sin." Let's break that down.

The "For whatsoever" piece means any *what*. The "not of" means against, opposite, or wrong. Lastly, "faith" would be my understanding (why) and experience (how). Therefore, sin would be any *what*, with a wrong *how/why*.

CHAPTER 12

Salvation Implications

NO DOUBT THE people that still hold to the "golden key" model will say the Bible states, "That if thou shalt confess with thy mouth the Lord Jesus, and shalt believe in thine heart that God hath raised him from the dead, thou shalt be saved" (Romans 10:9). There is no doubt this is true; however, we have to make sure this doesn't contradict the other "getting to heaven" verses. We have seen two so far: You are saved by grace through faith (Ephesians 2:8) and the Sermon on the Mount's "persecuted for righteousness' sake" (Matthew 5:10).

We've seen that saved by grace through faith is a process. What this means is that you have to continually grow in faith and grace in order to eventually be saved. Essentially, salvation is a process that depends on progress, not a quantitative threshold. What does this mean?

It means allowing God to influence our hearts and intentionally choosing to let Him direct our actions eventually saves us. How are we to know if the influence we get on our hearts is coming from God and not ourselves? We need to have faith because we can't see God and the effects haven't happened yet. We have to use our experience with, and understanding of God in order to determine if we are going to live by grace.

We know that progress is a result that depends on the individual's decision to pursue growth in grace and faith. This decision is an expression of the will of the individual and causes the individual to be saved when he dies.

Going back to our party analogy, when a person dies, it is up to God whether he gets to continue the journey or not. If people want to get to the party, they are going to choose to make progress toward the party instead of looking for short-term thrills by driving on roads that take them away from the party.

If the person should die while they are trying to make progress towards the party, the individual is saved. The proof he wanted to take direction from God (his heart) was in his actions. God allows him to continue the journey because it is in line with the will of the individual when he dies. It would be unholy for God to do something against the will of the individual. The true will of the individual is shown in his intended actions.

If people die while they are intentionally driving away from the party, then they are not saved, even if they said they wanted to go to the party while they were driving away from it. Even if they have spent their entire life driving toward the party, if they are choosing to drive away when they die, God will not allow them to continue the journey. Their will is showing the belief in their heart; they'd rather be comfortable in the moment than get closer to God.

Notice that the rate of progress isn't important to salvation. As long as you are moving forward when you die, no matter how slowly, it would be unholy for God to bar you from the continuation of the journey. Remember the servant with one talent would have been okay if he had simply obtained interest on the talent.

Getting back to the original verse (Romans 10:9), it would seem that "believe in your heart" is the important phrase. We've already seen this can't simply be a feeling. That wouldn't be holy. The "golden key" people want to believe this is a one-time event. They had a good feeling after they made their confession and now they are saved. This is contradictory to verses that show salvation as a process. It is contradictory to the concept of a free will. Clearly, believing in your heart is proven by actions, just like faith and grace.

We are saved by causes. "Believe in your heart" must be a cause that leads to an effect. Progress during the salvation process is the effect and is proven by actions. There are several passages that show what is in our hearts is proven by our actions, not what we say. One example is:

> "And they come unto thee as the people cometh, and they sit before thee as my people, and they hear thy words, but they will not do them: for with their mouth they shew much love, but their heart goeth after their covetousness" (Ezekiel 33:31).

Salvation Summary

There is a cause (heart) and an effect (action)—it must be intentional. God can't take people to heaven against their will; it would be unholy. People need to express their will, not just by saying it, but also by doing it. They need to grow in grace and faith (Figure 5).

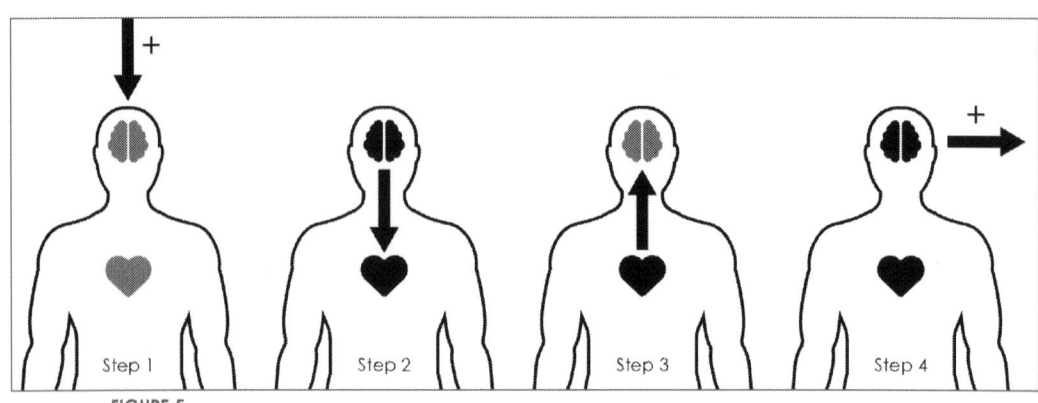

FIGURE 5

The first step involves God's divine influence on the individual (Step 1). This giving of a value to the individual is the first half of the biblical definition of grace. Remember, God supplied the value each servant was given in the parable of the talents. It is a gift. The next step is for the individual to choose to let this influence affect their heart (Step 2). They can close their hearts to this divine influence and be influenced by their flesh or another source. It takes faith to let an unseen God affect our lives. This ability and trust to open up your heart to God can be improved with more experience with and understanding of God.

Notice, God can influence our hearts without going through the first two steps. This four-step process shows a person focused on intentionally working out their salvation. God can give us a value while our focus is on other things and we may find ourselves suddenly at Step 3.

The third step in our intentional process is for the individual to check what God has said to do (Step 3). The individual is using his or her faith, checking the influence against his or her experience with and understanding of God. Steps 2 and 3 are consistent with our Faith Model.

The final step is for the individual to choose to have this influence come out in their actions (Step 4). This is the other half of the biblical definition of grace. Notice this results in a value. The value doesn't have to be equal to the value God first gave to the individual. However, like in the parable of the talents, people who choose to use the value God has given them will be able to return some value in addition to what God first gave them.

Also, notice the original value (Step 1) given to the individual was a "gift" that is in their control. This gift of grace is always being given to the individual. Like any gift, the person receiving the gift determines the value. They can do nothing with it or they can choose to receive it and use it; however, not requiring a return on this gift violates justice.

Looking at this process, the individual can become better if they improve their ability to receive the influence and let it come out in their actions (grace). They can also become better by improving their ability to determine (faith) if the influence (grace) is coming from God or another source. Both of these abilities are improved with more experience and understanding. Finally, since we'll never have physical proof of salvation, the entire process has to be by faith.

Grace Only

There are people who think they can be saved or grow by grace without faith. This manifests itself in two forms. First, there are people who do what they believe God is telling them to do without checking it against faith (Step 3). If all we had was grace (Steps 1 and 4), then we would be wrong. We make mistakes. We will misunderstand what God said (Step 1) or that what we heard was from our flesh and not from God.

Fortunately, we have faith to help us (Steps 2 and 3). Faith is built through understanding and experience. When we get an influence on our heart, we need to check if it is divine by using our or others' understanding and experience before we let it come out in our actions (Step 3). As one grows in faith, this process of validation becomes quicker and easier.

The second form is when people do whatever they want, thinking that no one sees them. Faith believes in things we cannot see. These people don't have faith in God, regardless of what they say. They essentially believe that because they can't see God, then God can't see them. This causes them not to open their hearts to God (Step 2). These people are giving their physical selves preeminence over their spiritual selves. In reality, they don't believe God really exists, and therefore, they can't be saved. Faith would remind the person that God is very real, even though we don't see Him, and consequences are very real, even though they haven't happened yet.

I'm sure there are still some among us who are uncomfortable with this book. They may even point to the verse (Acts 4:12) that says we must be saved under "none other name" except Jesus. I would agree with that, but I'd also point out what Jesus said in the Sermon on the Mount about getting into heaven ("persecuted for righteousness' sake"). There is no mention of His name.

All of these passages are true and say the same thing. If we were to put a name to the process, it would be Jesus, because He made it possible for us to receive the value. There is "none other name" that can be correctly placed on this process. However, a person doesn't have to know the name of Jesus in order to receive the value. The quoted verse that opens this chapter does not say it is the only method. It says it is a sure way of making it to heaven. One could even say it is an intentional way of eventually getting saved.

Righteousness, justice, faith, grace, confession, and repentance are concepts everyone can understand. Everyone can benefit from these causes regardless of their religious affiliation. It would be unholy for someone to intentionally make progress and be barred from continuing the journey because they didn't recognize the brand name on the process. In fact, the "golden key" belief in the name only leads to the biggest problem with the salvation process.

Evangelism

This has implications on how people help others get saved. The problem with the "golden key" definition of grace is that the ability to save people is completely out of our hands. In some cases, the convert has no control over their own salvation. This violates free will and the concept of what is holy.

Actually, there are two steps to intentionally help people get saved. People need to be aware of the process; then they need to do the process. Making them aware of the salvation process is evangelism. Helping them do the process of salvation involves teaching. The skills for each step are completely different.

Evangelists are like salesmen. They find people and try to convert them through preaching. Their goal is to motivate people to begin the salvation process. This is accomplished when a person expresses their will to emphasize the spiritual self over the physical self. Since only one self can take precedence, they are essentially killing the physical self and birthing the spiritual self. This is the event Jesus spoke of when he said we must be born again (John 3:3).

This process is a one-time event. Consequently, their approach usually emphasizes emotion. In fact, unless they can explain God and salvation in a logical fashion, they must make their appeal based on emotions and knowledge, which results in pride. Evangelists tend to focus on appearances and use all the techniques of a salesman. They don't have to worry about contradicting themselves because once they make the person aware of the process, they are gone.

Teachers are like farmers. They do their job every day. Their goal is to help the same people continue the salvation process. This process is a continual event. Consequently, their approach emphasizes thinking and understanding. They have to consistently help people grow. Therefore, the teachers need to be able to build relationships and not contradict themselves because they are going to be in front of the convert for a long period of time.

It would seem that we only need 5% evangelists and 95% teachers, yet when you look at the church, these numbers are reversed. Is this because evangelism is easier? Is it because people don't understand what to do well enough to teach others? Is it because people think they are going to get rewarded for getting people to pray one prayer and abandoning them?

Have you ever bought a product from a salesman, only to realize much later, after you thought about the sales pitch, that he didn't make sense? By then, it was too late. You were stuck with your purchase and the salesman got his commission. There are people who look at evangelism the same way. As long as they make the one-time sale, they believe the customer is stuck with the purchase and they are getting a commission from God.

Jesus didn't dress like a salesman. On the contrary, His actions often drove people away. He was interested in teaching. In fact, the Gospels use some form of the word "teach" sixty times, while some form of the word "preach" is mentioned thirty-eight times. Also, we see the conflict between salesmen and farmers after Jesus. For example, Paul continued to focus on teaching by revisiting churches he founded instead of starting new churches. Paul focused on understanding, not emotions.

If I were working for the enemy, I would tell people I was a Christian and all they had to do was say they believed in Jesus. I would appeal to them emotionally and overload them with facts that made them feel prideful. I'd tell them the "golden key" concept—they were saved by unmerited favor no matter what they felt and how illogical it appeared.

First of all, this would cause the "convert" not to receive salvation because he would feel like he was growing in knowledge, but actually he wouldn't be growing in the understanding and grace of God. Essentially, I would fuel the "convert's" pride.

Second, this "convert" would be telling people who are unaware of God that this is what God is about. He would be such a poor advertisement for Christianity that the people who are unaware would have no interest in understanding more. They would realize this approach doesn't make sense and is based on emotions. They would realize that "Christianity" is a shallow belief system for simpletons. This would cause the unaware not to receive salvation either.

Evangelism is necessary, but teaching is the solution to strengthening the "brand" of Christianity and helping more people make it to heaven, and eventually, the party. It would seem that we should be focusing on helping others deepen their experience with and understanding of God. In effect, we should be focused on quality rather than quantity.

- If you died today, would you be saved?
- How can a one-time prayer prove you will always listen to and obey God?
- What would the actions of a person who is choosing not to listen to or obey God look like?
- What would the actions of a person acting without faith look like?
- The last time you witnessed to someone, did you commit to continuing support? Where are they now?

Joel Swokowski's *Commentary*

There's a lot of tradition in the church regarding the concept of "Once Saved, Always Saved." People want to know, can I lose my salvation? My follow-up question: Why does the answer to that question matter?

What ought to be our objective? I would hope it is to know the truth!

In the many debates I've had over this topic, I started to use the word "forfeit" instead of "lose" which has led to many more profitable discussions.

Either a person can forfeit their salvation or they can't.

What is the worst and best that can happen in either case if they are wrong?

If they *can't* forfeit their salvation, then the best thing that can happen is that they sin without going to hell.

If they *can't* forfeit their salvation, then the worst that can happen is *nothing*...the person who thinks they can forfeit their salvation doesn't do any damage by being wrong about this because they are still focused on righteousness and they don't hinder others.

If they *can* forfeit their salvation, then the best that can happen is that they help others avoid hell.

If they *can* forfeit their salvation, then the worst that can happen is that they lose their salvation by being a stumblingblock to others by telling them they can't forfeit their salvation. If this person believes they can't forfeit their salvation and their focus is to be right instead of being contrastive and considering they are wrong, they are in danger of forfeiting their salvation.

My goal here is to help people who believe in "once saved, always saved" *consider* that they could be wrong…and ultimately get the discussion in the right place:

"What is the measure of salvation?" This is the question we ought to be asking instead of "can I lose my salvation?"

The answer to "What is the measure of salvation?" is "how I respond to sin!"

If the measure for salvation was "not sinning," then no one would go to heaven. We all sin, even Christians (1 John 1:8). What separates a Christian from a non-Christian is their response to sin. Christians confess and repent!

The Name of Jesus

Do I need to say the name Jesus, or know the name Jesus for me to live by grace and faith?

What if I'm pronouncing Jesus' name wrong? Isn't it Jeshua? Yeshua? Joshua?

Does it have to be in Hebrew? Greek? Aramaic?

The point is, *what* we call Him isn't nearly as important as knowing who He is as a cause: Right and Just. If I'm unaware of what the road I'm traveling on is called, it doesn't change the fact that that road has a name. I live on Wallace Street in my small village in Wisconsin. I'd still be living on Wallace Street if I was never told what my street's name was. Christ is the only way (road) to the Father. If I take a road to the Father without knowing what the name of that road is, it doesn't diminish the fact that Christ paved the way!

CHAPTER 13

Salvation and Rewards

SALVATION DEPENDS ON the reason causing the action, or, "Doing the right things for the right reasons." We can now look at the complete Salvation Model.

Salvation Model
1. Spiritual focus instead of physical focus
2. Dependent on the expression of your will to follow God
3. Your will at the time of physical death determines getting saved
4. The process begins with a divine influence (gift) that you allow
5. The influence is dependent on your uniqueness
6. Faith tests whether the influence (intent) is from God or not
7. The process is proven by intentional actions
8. The ability to follow God is built through experience and/or understanding
9. The entire process is solely between you and God
10. Mistakes are repaired with confession and repentance to God

Salvation is not a one-time event. It begins with a commitment to give the spiritual self preeminence over the physical self. After this "born again" event, salvation is a process that we need to continually grow in. We need to express our will to God that we want to take direction from Him instead of ourselves. We need to continue to become better at hearing God and letting it come out in our actions (grace). We need to continually grow in our experience and understanding of God (faith). This is more consistent with the idea of what is holy, in that salvation isn't dependent on everyone crossing the same line; it is dependent on the individual, what they were given and what their will is, as expressed in their actions—just like the parable of the talents.

The "golden key" explanation for salvation is represented in Figure 6. Panel 1 shows a person who has taken a value away from another person. Person A now has a value and Person B is missing a value. Justice would say God needs to equal the situation (Panel 2). That is, God needs to take a value away from A (punish) and give a value to B. (However, we saw that this can only be done once Person B forgives Person A. Otherwise, Person B could take a value from Person A through revenge or speaking ill about them.)

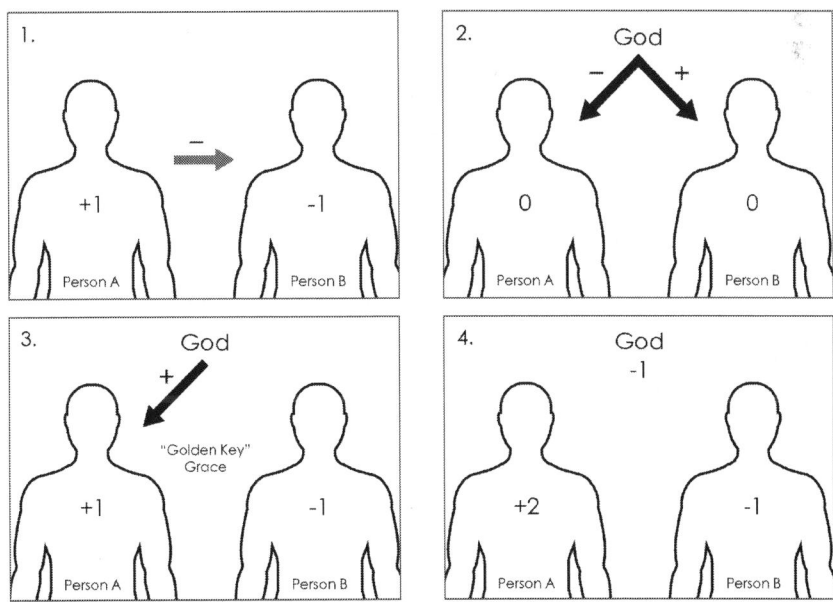

FIGURE 6

However, "golden key" grace doesn't agree with justice. Panel 3 shows "golden key" grace. God gives the person a value in order to cover their sin and allow them to go to heaven. The result is shown in Panel 4. God and Person B are out a value, while Person A is further ahead.

Notice this process doesn't involve thinking. In fact, looking at it in this fashion shows why "golden key" faith requires believing in things that don't make sense. "Golden key" grace tries to account for salvation and rewards. Remember, the Bible speaks of salvation and rewards as two separate processes. Jesus spoke of them separately during the Sermon on the Mount as well as in His parables. Salvation is qualitative, while rewards are quantitative.

We will need two separate models to explain salvation and rewards; otherwise, we are going to end up with one model that doesn't explain either process well. Remember, justice says everyone will have to pay for the wrong they've done, even people who are saved!

Reward Model

At this point in the book, we are still in the process of identifying the Reward Model. So far, the Model looks like this:

When we do something unholy to someone else (Panel 1 of Figure 7), justice says we (Person A) deserve punishment (Panel 2). God could punish us immediately, but mercy allows God to put off immediate punishment. We know we have been unholy (sinned) because of guilt. What we decide to do with our guilt is an expression of our will.

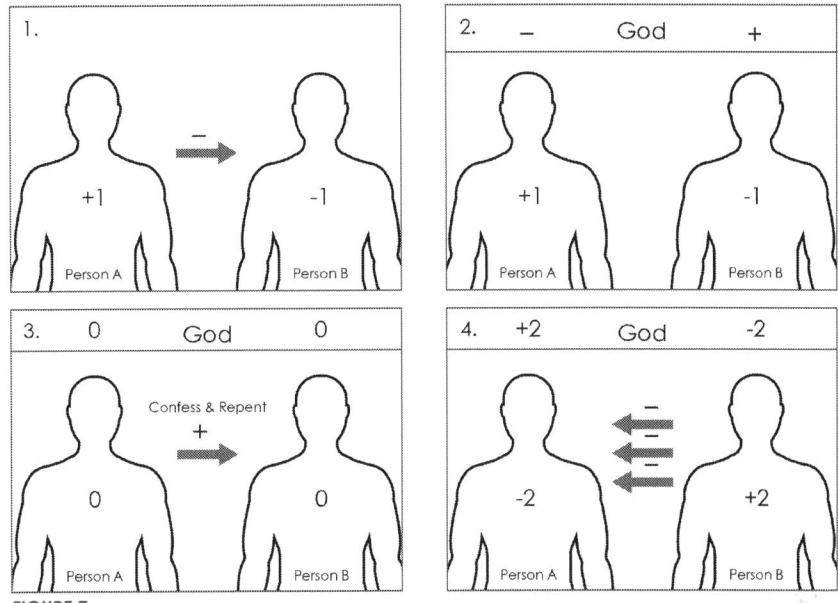
FIGURE 7

We have the opportunity to pay for our lack of holiness and prevent the punishment by making it unjust (Panel 3). We can remove the guilt through confession and repentance. Confession and repentance can also lessen or help us avoid punishment because it is the process that makes us holier. (In the Salvation Model, God counts our decision to become more righteous as a value and we receive a portion of the value Jesus acquired to pay for our sins. The Reward Model deals with our interactions with people and is driven by justice).

If we decide we aren't going to get punished and proceed to do more unholy acts, at this point, God is being merciful when He punishes us. This punishment is intended to stop us from incurring further debt.

Notice, it is possible for us to be owed a value if the offended person (Person B) doesn't forgive us and continues to take a value from us (Panel 4). Even though we initiated the conflict by taking one value, we can end up being owed because the wronged person takes three values! This also begins to explain one of the reasons why "bad" people prosper in this life. The answer? Because people don't forgive them!

Do you remember the non-contradictory definition of forgiveness? Forgiveness is when the individual states his will that he won't attempt to get his own justice even if he has the opportunity. Once God is sure the offended individual won't work to get his own justice, justice can occur.

The example given above is just the beginning of the Reward Model. The actual model would have to account for all the people watching Person A interact with Person B. These people can resent Person A for what he did to Person B, even though they weren't involved in the exchange of value. If all of these people continue to resent Person A and speak ill of him, then Person A will accumulate value from all of these people as well!

Notice that every interaction with another individual is an exchange of value. If you understand the Reward Model, you can quantify the gains and losses for every individual resulting from every interaction that occurs.

People know they should forgive and love each other. However, knowing this doesn't result in people forgiving and loving each other. Unlike the Salvation Model, people can get the effect without understanding. When people do understand the Reward Model, they eventually forgive people more quickly and love others more, but it seems to be a three-stage process.

The first stage occurs when the individual realizes during a stressful situation that if they retaliate they are going to lose value. This thought actually stops them from retaliating. After the moment has passed, I usually get a call telling me how they, "didn't lose a value today," when they normally would have. I also point out to them how they also don't have the guilt they would normally have if they had retaliated.

The second stage occurs when the individual realizes before a stressful situation that if he retaliates he is going to lose value. For example, a good friend of mine was notified on the way to a job site that his company had failed to provide a covering to keep him out of the rain, but he still needed to complete the job. He called me on the way to work to tell me every drop of rain that hit him was only going to be a value for him because he wasn't going to cause a scene.

The third stage occurs when the person gives people a value without thinking about what they will get out of it. This is when they are operating in grace.

I've had numerous people tell me it shouldn't be like this. People should just want to forgive and love without getting anything back. However, that method has been used for years and it doesn't work. This method works and is based on the Bible. Remember, Jesus motivated us to love our enemies by telling us we would get a reward. Jesus helped us understand. People who fight this method are really fighting a technique Jesus used.

Two Models

The Reward Model actually says that people can get a value for doing the right thing for the wrong reasons because of justice. The Reward Model accounts for works. The Salvation Model is based on doing the right thing (righteousness) for the right reason (because God specifically told you). Now we truly see why "golden key" salvation is contradictory—it tries to account for two models (Salvation and Reward) that are separate and contradictory.

Jesus actually had two messages: 1) How to get to heaven and 2) How to get rewards on earth and in heaven. It seems every book about Jesus (except for the Bible) tries to resolve these two messages into one by ignoring the verses connected with the other message, abstracting Christ's words until they apply to both, or arguing for interpretation/translation errors.

The first message is the Salvation Model, which is qualitative. Whenever Jesus said that it doesn't matter how long you have been doing the process and everyone gets the same value, He was talking about the Salvation Model. The Salvation Model is referenced in verses that show God and not ourselves as the source of earning a value. All we do is choose to obey.

The second message is the Reward Model. Whenever Jesus spoke of works that lead to rewards, He was referencing the Reward Model. These verses show everyone ends up with different value. Clearly, these are two separate models. Some books focus on our ability to do works. Other books focus on our inability to do anything when it comes to interacting with God. The need to make these two messages into one can lead to some alarming results.

When people or other religions say we have to love our enemy, tithe, suffer persecution, etc. in order to prove we are saved, they are basing salvation on works. This is legalism. Instead, these works lead to rewards and have nothing to do with salvation; otherwise, salvation can be earned. After all, reward means "to recompense both good and bad." This sounds like justice; now we see why the Reward Model is quantitative and relies on justice.

Most people reconcile the contradiction by simply ignoring the Reward Model verses. Another reason these verses are ignored is they are unpleasant to our flesh because the only ways to get positive rewards are: 1) Give a value and don't get anything in return and 2) Handle unjust suffering well. Both of these events run counter to a message that Jesus came so that we can be happy, successful, and problem-free without thinking. In fact, the New Testament consistently says we should rejoice in tribulation because it is an opportunity to gain a value.

Furthermore, Revelation 20:11-15 describes two judgments:

> "And I saw a great white throne, and him that sat on it, from whose face the earth and the heaven fled away; and there was found no place for them. And I saw the dead, small and great, stand before God; and the books were opened: and another book was opened, which is the book of life: and the dead were judged out of those things which were written in the books, according to their works. And the sea gave up the dead which were in it; and death and hell delivered up the dead which were in them: and they were judged every man according to their works. And death and hell were cast into the lake of fire. This is the second death. And whosoever was not found written in the book of life was cast into the lake of fire" (Revelation 20:11-15).

The first judgment involved books and was according to works. It would certainly take "books" to list all the works of every person who ever existed. This is the Reward judgment. The second judgment involved one book (book of life) and determined salvation. It would only take one book to list just the names of the people who are saved. This is the Salvation judgment.

The existence of two messages is best shown when Jesus was asked which commandment is the greatest. The question required Jesus to name one commandment. His response?

> "And Jesus answered him, The first of all the commandments is, Hear, O Israel; The Lord our God is one Lord: And thou shalt love the Lord thy God with all thy heart, and with all thy soul, and with all thy mind, and with all thy strength: this is the first commandment. And the second is like, namely this, Thou shalt love thy neighbour as thyself. There is none other commandment greater than these" (Mark 12:29–31).

Jesus was making it very clear he had two separate messages, not one. The first commandment is the ultimate example of the Salvation Model. The second commandment is the ultimate example of the Reward Model. In the next book, we will define love and see that love is the ultimate result of God being righteous and just. Consequently, the ultimate examples of both models would be an effect of love.

Notice again, the Salvation Model relies heavily on righteousness, while the Reward model relies on justice. That's because God, who is righteous and just, created these two models.

Throughout this book, we have seen separate verses for salvation and reward. Now that we know the Salvation Model doesn't cover the reward verses, we must uncover a Reward Model to account for everything Jesus was saying. Before we present the beginnings of the Reward Model, let's look at one last difference between it and the Salvation Model.

The Reward Model is separate from salvation in that our issue is with another person, not God. Remember, salvation is solely between God and us. Notice that confession and repentance also work in this Reward Model; however, the focus of our confession and repentance is on another person for the incident instead of towards God for not following grace.

The Reward Model says that confessing and repenting to God has no effect on resolving the inequity between us and another person. In order to resolve this issue, the confession and repentance have to be towards the offended person.

Reward Model
1. If a person gives a value, justice equals it out in spiritual value
2. If a person takes a value, justice equals it out in spiritual value
3. God won't deliver a value to you until you forgive
4. Doing the right thing is rewarded, even if it is for the wrong reason

- When is the last time you confessed and repented to God for not following grace?
- When is the last time you confessed and repented to another person for the value you took?
- Describe the last time you forgave someone. Do you still feel like that person owes you?
- Take one of the previously mentioned salvation verses and describe how it relies on being righteous.
- Take one of the previously mentioned reward verses and describe how it relies on justice.

Joel Swokowski's *Commentary*

This chapter starts with this statement, "Salvation depends on the reason causing the action, or, 'Doing the right things for the right reasons.'" Another way of saying this is, "Salvation depends on doing the right *what* with a right *how/why*."

This chapter also gives the key to understanding Jesus' parables. Jesus had two messages:

1. How to receive salvation.
2. How to gain reward.

Jesus' parables are best understood when you first determine if it is a salvation-parable or a reward-parable (or possibly both). If the parable is talking about the same benefit that everyone receives, it's a salvation-parable. If the parable is talking about different levels of benefit depending on merit, it's a reward-parable. Salvation is qualitative, reward is quantitative.

The key difference in the Salvation and Reward Models is in who the cause of the interactions are: Salvation is our interaction with God, and Reward is our interactions with people. The implication: Jesus' death on the cross brings salvation through the forgiveness of sins that we commit against God, while the sins we commit against people affect our reward and are not covered by Jesus' blood. This is why Jesus spent time teaching people how to deal with the trespasses we commit against each other (Matthew 18:15-17, Luke 17:3). This also means that when I'm confessing and repenting to the Father for the sins I commit against a person, I'm not following the words of Jesus.

In fact, Dietrich Bonhoeffer wrote in "Life Together" that confessing to God for the sins we do to others is really a confession and repentance to ourselves (which would be idolatry). If a person truly did confess and repent to God for the sins they did to another, God would tell that person to then go and confess and repent to the person.

Forgiveness

Forgiveness was defined in this chapter as "a statement of my will that I won't attempt to equal out my own justice even if I had the opportunity." The end of that statement ("...even if I had the opportunity") helps people really understand that forgiveness means I'm stating that I'll *never* take out my own justice here, in *any* context. I've also found that the ultimate measure for when I've forgiven someone is when I've not only committed to not equaling out my own justice but I'm also committed to helping the person who's been unjust towards me.

The first stage of growth in the realization of this model is to say, "It's all about justice." It is a great first step. However, I know people really get this model when they then say, "It's all about grace." When you realize the power of justice, you are one step away from getting it. Don't be in a rush.

In the Apostle John's First Epistle, it says there is a sin "not unto death." Therefore, there are three types of sins:

1. Blasphemy of the Holy Spirit — The unforgivable sin that affects Salvation (Matthew 12:31-32).
2. Sins against God — This affects Salvation ("sin unto death") and is forgivable.
3. Sins against people — This affects Reward ("sin not unto death") and is able to be covered with love/repentance.

CHAPTER 14

The Way

THE SALVATION MODEL accounts for four seemingly contradictory aspects of salvation:

1. **Salvation can be obtained instantaneously regardless of the individual's past.**
   If the individual dies immediately after choosing to act according to faith and grace, regardless of their past, they will be saved. "Though your sins be as scarlet, they shall be white as snow; though they be red like crimson, they shall be as wool" (Isaiah 1:18).

2. **Salvation is a gift.**
   "For by grace are ye saved through faith; and not of yourselves: it is the gift of God" (Ephesians 2:8). Salvation begins as a gift because the first half of grace is a gift, but it is only half of grace. There may be nothing you can do to obtain this divine influence on your heart; however, the idea of holiness allows that what we do with this gift is still the responsibility of the individual.

3. **Salvation can be rejected.**
   The Bible clearly states people can fall away from the faith (Hebrews 6:6) or be removed from the faith (Matthew 18:17)

after they became a believer. Several parables describe favored servants of a master that make bad decisions and lose their position. Therefore, salvation must be something that can be rejected after the individual begins the process. We can't be saved against our will. If we choose to no longer follow grace (God's influence), then we are telling God we don't want to continue the journey. If we choose not to repair our sin with confession and repentance, then we are telling God we don't want to grow in righteousness.

4. Salvation is based on uniqueness.

Finally, salvation can't be dependent on some universal quantitative level of outward performance (works) that is the same for every individual. It wouldn't be holy for those who more naturally perform works to be closer to heaven. It must take into account the uniqueness of the individual and his intent. One of the places we saw this was with the parable of the talents. People begin the journey with different levels of gifts.

The Salvation Model presented in this book is the only explanation that doesn't contradict these four aspects of salvation, as well as the qualities of righteousness and justice. Every other explanation for salvation accentuates one or two of the above four aspects and falls short in at least one of the other requirements. Let's look at just one example for each of the four points.

Looking at the first point, there are people who have problems with murderers getting saved because of a deathbed confession. However, the Bible is clear that it doesn't matter how bad a person's past was—everyone can obtain salvation. This is the good news! As for point #2, some people think this means there is nothing we can do about our salvation. While the process of salvation is something that is given by God, holiness says we must be responsible for the result. Otherwise, point #3 would not be true. Finally, believing that works saves us is a denial of #4.

Sanctification

There are still those presenting salvation as a one-time event. Their problem is they have no way to motivate people to grow in faith and grace. The way they address this issue is to introduce the process of sanctification to the believer. Sanctification means "to set apart." Basically, the believer is told they must continually strive to grow in holiness and set themselves apart from the world, even though they've already been saved.

When the believer asks for the value in doing this, the response is abstract. The Bible supports the idea of growing in holiness; however, in the context of "golden key" salvation, there is no sense of urgency to participate in sanctification. After all, we will never become perfectly holy. What's the benefit of participating in a process if the individual is going to receive the same ultimate benefit as both someone who doesn't participate and someone who has made tremendous progress through the process?

The Salvation Model says the salvation process is part of sanctification. Attaining a universally accepted quantitative level of performance is not the goal. Progress is the goal. Progress, no matter how small, objectively proves your will to take direction from God and continue the journey toward taking increasingly more direction from God once you die.

The value of sanctification is that it ensures your salvation. The more progress you make, the less likely you are to be permanently swayed by momentary trials and temptations. Let's take this opportunity to look more closely at how people can lose their salvation.

When a person does something, there is a period shortly after the decision where the person may not be sure if the act is apart from grace. I refer to this period as the "mercy period," because the individual is still under mercy. They don't know for certain the act was apart from grace. Remember, we are saved by grace through faith, so we must be given time to check the action against understanding and experience. Salvation is lost when the person knowingly chooses to act apart from faith and/or grace. Choosing to not think is the same as acting apart from faith. With grace, the person knows it is apart from grace because he experiences guilt immediately.

The individual will know the act is sin when he experiences guilt. It is at this point the individual expresses his will. Ideally, the person would remove the guilt by confessing and repenting. In this situation, the person is never in danger of losing his salvation. Holiness says he could have died at any time during this situation and God would have allowed him to continue the journey. Furthermore, repentance would result in a modification of faith. The individual now knows this action results in sin. This should increase his faith and make a future occurrence in this area less likely; however, if he chooses to ignore the guilt, his conscience begins to become less sensitive and the guilt will turn into pain. When people experience pain, they usually get rid of it by unloading the pain onto others. This unloading of pain is really a cry for help.

A mature Christian is able to recognize this behavior and bear the person's pain in an attempt to help him or her remove the guilt that causes the pain. If the person doesn't unload the pain, it will turn into fear. Also, it is possible to sear over the conscience to the point that he or she is unable to feel guilt. The Bible says in 1 Timothy 4:1-2 that these people with a seared conscience have departed from the faith. We know these people will not be saved if they die, but at what point did they lose their ability to be saved?

If the person would die during the time they recognized the act was sin and before they confessed and repented, it would depend on God's mercy. How much mercy did the person have? People who aren't merciful get little mercy from God. Would they have confessed and repented before their mercy ran out? Only God knows. It would seem unholy for a person to lose their salvation because they happened to die during this period while they were still under mercy, especially if their death was something they didn't cause.

In order to improve one's ability to achieve salvation, it seems the individual would maintain a sensitive conscience and give as much mercy to others as possible. Improving these attributes is both an effect of the sanctification process. Furthermore, sanctification means "set apart." It can best be explained by going back to the party analogy. Imagine that when you got directions to the party, you were in a crowd of people. Your decision to emphasize your spiritual self over your physical self was the same as deciding to leave the crowd and make progress towards the party. It is impossible for you to make progress with others. You are unique. Not only did you leave the crowd at a unique time, but you also have a unique rate of progress. The only way for you to make this journey with someone else is if you both move at exactly the same rate. Said another way, the only way to be with others is to not make progress.

Since uniqueness proves that your progress can't be exactly the same as the rate of progress of others, your journey is made alone. You are "set apart" from the crowd and every other person on their journey to the party. The choice to be set apart through progress is sanctification.

Jesus Model

The fact that we will never be perfect during this existence leads to another subtle misconception that results in a barrier to the salvation process. The Bible tells us "all have sinned" (Romans 3:23); however, Jesus never sinned. Everything He did was righteous and just. Remember the model presented in Chapter Two?

God Model	Jesus Model	Human Model
1. Righteous	1. Righteous	1. NOT righteous
2. Just	2. Just	2. NOT just
3. NO free will	3. Free will	3. Free will

Since Jesus never sinned, it is obvious that we humans don't fit the Jesus Model. Holding others and ourselves to the Jesus Model will only result in frustration due to lack of progress. While Jesus is our standard, we will never be able to perfectly emulate Jesus during this existence. The process of emulating Jesus is called "discipleship."

In order to avoid frustration during this existence because of unrealistic expectations, it looks like we need to emulate someone who fits the Human Model. Our examples from the Human Model are the disciples. In fact, there is one disciple who wrote more than anyone about his discipleship journey.

- What could you say to someone who believes his or her past is too bad to be saved?
- What could you say to someone who believes they have earned their salvation?
- What could you say to someone who believes they can't lose their salvation?

- What could you say to someone who believes everyone must look and act the same in order to be saved?
- In what areas can you work on to become more holy?

Joel Swokowski's *Commentary*

Have you ever asked your pastor (or any Christian for that matter), "what do I do after salvation?" The question seems as though it should be easy to answer *unless* you believe that all Christianity is about is getting saved.

Sanctification was defined as "to set apart." It was also described as a growth in holiness. How do we do this? I love the answer to this question, as it shows one of the many ways in which God has given us simple instructions… simple, not easy! To grow in holiness and in sanctification only requires a person to grow in the same causes that saved them: grace and faith. So, grace and faith save me, and my growth in grace and faith sanctifies me! This also means that my assurance of salvation comes through my growth in sanctification!

Sanctification is a benefit! Why don't people see it as a benefit? They either don't understand it or they want to do their own plan, not God's plan.

So, what's the answer to "What do I do after salvation?" Grow in sanctification!

This chapter also dives deeper into the concept of "losing" my salvation. Now I'd once again recommend using the word "forfeit" in the place of "lost," especially if you're having a tough time with this topic.

I like this chapter in that it expands on the benefit of mercy, especially the mercy we experience from God. We've seen that mercy is the "postponement of judgment." But God isn't merely delaying judgment on people for the sake of delaying judgment. There is a greater purpose to mercy: the time between the injustice and the judgment is meant to give you the time to make up for the injustice. How are you handling the mercy periods that you're experiencing?

CHAPTER 15

Discipleship

PAUL WROTE MORE than any other biblical author about the process of discipleship. There are four key characteristics that distinguish Paul from Jesus:

1. **Paul's admission of a sinful past.**
   Acts 22:4 says Paul admitted that he persecuted Christians unto death.

2. **Paul had a conversion experience.**
   Paul's conversion (Acts 9:3–6) was a supernatural event. The process of salvation is incomplete without a supernatural interaction. This book focuses on proving a non-contradictory explanation of God and salvation. This is an intellectual explanation.

   Since God is a spirit, our ability to do this process depends on our spiritual response. Knowing this process and doing it are two separate events. It is possible to do this process without completely understanding it. However, the more one understands the process, the better they are able to intentionally make progress towards God, and to help others do the same.

Conversely, depending completely on salvation by solely knowing the process and not doing it is called "mental assent." Salvation cannot be obtained this way because there is no spiritual connection. We've seen "believe in your heart" is the key because it leads to doing the process.

3. **Paul saw salvation as a process.**
In Philippians 2:12-13, Paul writes, "Work out your own salvation with fear and trembling. For it is God which worketh in you." Paul sees salvation as something in which we have control; however, it is ultimately God who does all the work. Our responsibility is to choose to let Him work through us (grace). This is consistent with both our free will and salvation being a gift of God. This process focuses on becoming more righteous. Notice, Jesus didn't increase His righteousness while He was on earth. Jesus personally worked the Reward Model to obtain the value we need to cover our sins. Another proof of salvation being a process is that Paul was always looking forward. In Philippians 3:13, Paul writes: "Brethren, I count not myself to have apprehended: but this one thing I do, forgetting those things which are behind, and reaching forth unto those things which are before." Paul doesn't see salvation as a one-time event. He sees it as something that will always be ahead of him during this existence and is not limited by his past.

4. **Paul was very open about not being perfect (Philippians 3:12).**
Among the numerous examples where Paul told us about his imperfection, one of my favorites is 2 Corinthians 12:7–9. Paul tells us he had a "thorn in the flesh" lest he should exalt himself in pride over the abundance of revelations God had given him. What was Paul's solution to dealing with the thorn? Paul asked God to remove it; however, God had another plan.

Paul says three times God's response was, "My grace is sufficient for thee: for my strength is made perfect in weakness." Basically, God is telling Paul he will get further ahead if he focuses on hearing from God and obeying instead of pursuing a solution in his own strength. Paul is showing us he is imperfect and will only attain perfection through his weaknesses, because he will have to live by God's influence (grace).

The story also serves as an excellent summary for the non-contradictory definition of grace. Notice Paul got the thorn because he had received a great revelation and others may exalt him excessively. If grace is "unmerited favor" needed to cover sin, what was Paul's sin? Paul didn't get a thorn because of sin. The thorn was so he would obtain the maximum reward in heaven rather than get it on earth by being excessively exalted. God gave Paul the thorn specifically so that others didn't see Paul as perfect. God's grace resulted in a better plan.

In Romans 6, when Paul is discussing grace and rhetorically asks whether we should sin so that grace abounds, he sums up the answer in verse 16. "Know ye not, that to whom ye yield yourselves servants to obey, his servants ye are to whom ye obey; whether of sin unto death, or of obedience unto righteousness" (Romans 6:16). This issue of obedience to grace is crucial.

Sacrifice

In this book, we have seen that faith and grace are causes of salvation. We have seen that good actions (works) are proof of salvation, but not the cause. The Salvation Model focuses on causes (faith and grace). The Reward Model focuses on results (works) regardless of the causes. However, these facts don't stop people from focusing on actions.

Remember, we are both physical and spiritual beings. Basically, our bad actions come from our physical self. These fleshly actions are short term and small picture. They are not holy and therefore, not from God.

One way to deal with these actions is to focus on actively stopping the bad act. This is called sacrifice. A concordance gives the definition of sacrifice as "to kill (animal) flesh"[7]. The focus is on killing the flesh that causes the sin. The belief is that if no bad actions come out, then all that is left are good actions. However, this doesn't mean the actions are for the right reason.

We have seen a second way to make good actions come out—grace. Instead of focusing on trying to kill our bad actions, we can focus on doing what God tells us to do—sacrifice focused on the physical (flesh). Grace focuses on the spirit and having a spiritual connection to God. Obedience to God is a positive measure that is attainable for us. Simply put, we can either focus on what not to do (sacrifice) or we can focus on what to do by obeying God.

The Bible says that to obey is better than sacrifice (1 Samuel 15:2). When we obey, we are doing good actions for the right reason. When we sacrifice, we are not doing bad actions, and whatever good actions result aren't necessarily for the right reason.

Another reason obedience is better than sacrifice is that focusing on killing our bad actions is a frustrating process because we will never avoid sin completely. We will never ultimately be successful if our measure is based on complete removal of sin. Remember, "All have sinned" (Romans 3:23), so we will never be free of sin. However, we can be perfect in our experiences with guilt (the result of sin). We need to focus on our ability to remove guilt.

This book covers the only effective way to remove the effects of sin—confession and repentance. These are actions in our control. These are the keys to becoming righteous. These are the tools for sanctification. If people are focused on holding themselves and others to the Jesus Model, they will either end up in condemnation or pride. Condemnation occurs when the individual realizes they will never achieve this goal.

On the other hand, the only way to consider yourself successful with a goal you can't achieve is to focus on how much better you are compared to others. The individual must actively identify others who are worse than him. This is pride. These people are experts at finding the faults in others and reminding people of their past actions.

The end result of focusing on sacrifice is a two-tiered society. On the "top" are the prideful, while the "bottom" consists of the condemned. However, those focused on emulating Paul will be focused on God. They will be actively looking to obey God and be humble enough to repair the instances where they didn't follow God. These people will be experts at admitting when they are wrong and skillful when it comes to forgiving others. This book calls these people who pursue discipleship "Christians."

One of the quickest ways to detect these Christians is to correct them. They will be humble and receive what you have to say, even if you are wrong. The prideful people will not hear you, even if you are right. They will be very stubborn in their beliefs.

In fact, this brings up one final point from the Bible. Stubbornness is not a positive trait. Stubbornness is the opposite of sanctification (actively looking for mistakes in one's self and changing them). The result of sacrifice is a prideful focus on comparatively justifying one's position.

The result of obedience is humility, and actively looking contrastively for correction.

Christians should be the best at admitting they are wrong. However, it seems that currently, "Christians" are prideful, stubborn, and not teachable. Remember, the Bible says stubbornness is as iniquity and idolatry (1 Samuel 15:23).

Fundamental Christianity

We know that Jesus is the only way to God because we know the reasons why. Why doesn't everyone realize this? I believe the reason is due to the way we evangelize. I like to illustrate this issue with the "Henderson 4000."

I love to tell people how much I love my new Henderson 4000. I convey to them how it makes my life easier, gives me more free time, and makes me happy. When an audience member asks, "What is a Henderson 4000?" I respond by relating more of the benefits: it is convenient, it comes in many colors, it doesn't need a lot of maintenance, etc.

This exchange will continue until someone tells me they don't think a Henderson 4000 exists. At this point, I will ask them why they don't believe in a Henderson 4000. The typical response is, "Because you can't explain what it is!"

At this point, I like to establish two facts. First, I obtain agreement that the more I talked about the Henderson 4000 in an abstract fashion without understanding, the less likely they were to believe in it. Secondly, I show the only way I could get someone to believe in the existence of the Henderson 4000 would be on a purely emotional basis. It is at this point I ask the audience, "How is the Henderson 4000 any different from God and salvation?"

Currently, the majority of those who do believe in Christianity initially did so because they were emotionally drawn. This is because Christianity, along with every other religion, is presented in an emotional fashion. However, this is not righteous and just. God's true nature stands apart from every other belief system because it can be understood and explained objectively.

There are plenty of leaders who believe they are giving an objective explanation of Christianity, but their explanations are contradictory to other leaders who also believe they are giving an objective explanation. This would be analogous to two or more people quoting facts about the Henderson 4000 to an audience simultaneously, but each person's presentation would seem to contradict the other presentations. The result would be the same; audience members don't believe the Henderson 4000 exists because no two people can agree on what it is.

Finally, some try to prove Christianity by showing that other religions are more emotional. While other religions are more emotionally based, that presentation doesn't prove Christianity is the only explanation for God and salvation. It proves that it is the best explanation to date.

No matter which of these abstract techniques we use, the result is the same: The world sees the contradictions and dismisses Christianity as the same as other religions. Emotionally based beliefs have no place in logical discussions. This emotional presentation is not teaching. There is no understanding. However, this is the current state of evangelism.

What are the principles behind all of these abstract instructions? There are only two:

- Do what God is telling you to do (grace).
- Confess and repent when you don't follow grace.

This is Fundamental Christianity. Grace directs us, not ourselves. People in pride believe that they create their circumstances and that they are either mostly or completely the cause of their good actions. This book proves that the ultimate cause of all of our good acts is God through grace. This does not mean we have no control. We do have a responsibility, otherwise, we couldn't be judged. We can choose to allow God to influence our hearts and let it come out in our actions. However, this role is supportive of God's intention for our lives and, at most, would only allow us to claim less than 1% of the credit for the results.

Notice, we can't claim any credit for the process of salvation itself. We did nothing to earn the actual process or the fact that God offers it to us. In this manner, salvation is a gift from God to us. There is no support to believe this means we have no control or responsibility for the result of the salvation process.

On the other hand, those who think they are the ultimate source of their good actions are in pride and will have a hard time continuing to admit their sins and remove their guilt. Remember, the current level is not important. The goal is to increase your progress and the rate at which you are letting God direct your life. Pride makes you think you've apprehended salvation and/or perfection. Discipleship follows Paul's example and guards against believing you have apprehended it.

Discipleship is shown when people confess and repent. This is an admission that the individual understands he is not perfect. Recognition of his lack of perfection causes him to forgive others. Realizing he ultimately relies on God results in true humility. The individual is teachable and driven to understand more. Aren't these results consistent with the effects of Christianity? Even the world knows that these are the traits of true Christians.

In fact, the Bible says in Acts 11:25–27 that Antioch was the place where the disciples were first called Christians. This label came from non-Christians. The proof you are following Christ should come from non-believers. Jesus said they will know you by your fruit (Matthew 7:20). After all, non-believers are the first to tell us we are not Christians when we aren't humble, don't forgive others, or fail to show mercy. The world knows the characteristics of a Christian!

Ultimately, the measure of a true Christian is not their lack of sin and ability to appear perfect, but their willingness to admit their imperfections, repair their mistakes, forgive others, give mercy, and take increasingly more direction from God.

This book has explained the causes that lead to these fundamental results. More than that, we have seen why every other explanation for Christianity is contradictory.

We've seen that grace directs us. When we fail grace, we sin, which causes us to experience guilt. If we don't remove guilt, we experience pain. If we don't remove pain, we experience fear. We can look at the process in this fashion:

$$\text{Grace} \rightarrow \text{Sin} \rightarrow \text{Guilt} \rightarrow \text{Pain} \rightarrow \text{Fear}$$

We have control over grace and removing guilt. Both processes are derived from righteousness. It would seem that we should focus our efforts towards improving our ability to do what God is telling us to do and remove the guilt when we don't follow grace. We called this Fundamental Christianity.

However, most Christians focus on dealing with sin, pain, and fear. Think about it. How many sermons have you heard explaining how to better hear from God and remove guilt? How many sermons have you heard about sin, pain, and fear?

- Describe your sinful past. Do you believe you have been forgiven?
- Describe your conversion experience. How did God interact with you?
- How have you made progress in your salvation?
- How would you handle the expectations of perfection placed on you by others?
- How have your God and Salvation Models changed since you began reading this book?

Joel Swokowski's Commentary

This book is like a textbook. It is filled with much information. Yet, Lenhart was explicit in his explanation of salvation in that we can never "know" enough to get to heaven. Salvation requires us to **do** the word and will of God. Lenhart explicitly stated that "mental assent" is not a doctrine of God. Then why has he still been accused of believing such? How many of the people who've come against the author and the information in this book have even read this book?

Yet again, we also see the resolution between credit and responsibility as it relates to our salvation: humans are responsible, and God gets the credit. Our responsibility in the matter of salvation is easily expressed in the explanation of "Fundamental Christianity":

1. Do what God is telling you to do.
2. Confess and repent when you don't.

Discipleship can be seen as helping others get to a point where they are living out Fundamental Christianity to a point where they can teach others how to teach others the same! Discipleship is replication. Replication of what? Replication of a person who can intentionally grow in holiness (sanctification) and help others intentionally replicate this ability into others, including the ability to replicate! The making of disciples of all nations is what Christians refer to as the "Great Commission" (Matthew 28:19-20). This is the job Jesus left us with and the job I believe He'll be asking us about upon His return.

Here are two verses to support this claim that Discipleship is replication:

1. John 15:8 — "Herein is my Father glorified, that ye bear much fruit; so shall ye be my disciples."

"Bear much fruit" means creating disciples. It's not enough for an apple tree to produce fruit. It needs to produce fruit that results in replication (reproduction). The offspring (reproduction) of a human is another human. The offspring of an apple tree is another apple tree. The offspring of a disciple is another disciple, not merely another "saved" person.

2. Luke 6:40 — "The disciple is not above his master: but every one that is perfect shall be as his master."

Notice, the teacher is making disciples and ought to be like his teacher, which in the end, makes more disciples.

Replication is how this Great Commission can go on indefinitely.

Replication ought to lead to generations becoming better over time.

Jesus did this. He replicated His **thought process** (the mind of Christ) into others and even said we'd do greater things than even He did (John 14:12)… **if** we are successful at discipleship!

Becoming Like Christ means the "mind of Christ," which means we empty ourselves (Philippians 2) and take direction from the Father, as Christ did. This results in us becoming more holy over time (sanctified).

One major issue with people not understanding what to do after salvation is that discipleship is not occurring. People with the belief that being saved is all we are meant to do as Christians are bound to miss out on growing in sanctification and helping others do the same. If all I'm focused on is making myself feel like I'm saved and maybe even helping others get saved, how am I "having the mind of Christ"? How am I helping others grow into having the "mind of Christ"?

Discipleship is an umbrella under which everything that God wants for us falls, from pre-conversion to leadership. Discipleship is the highest level of sanctification and therefore, discipleship is the ultimate action for a Christian to take after they become saved.

Why has this book helped so many people intentionally become and make disciples? Doctrine.

Doctrine is the key to discipleship because replication needs doctrine! Doctrine is the contextless answer that *anyone* can use in *any* situation. Discipleship accounts for a person following grace in their uniqueness. Without doctrine, I can't help you become more like Christ (having the mind of Christ), I can only help you become more like me. Doctrine means that I don't need to memorize every situation (apologetics). With doctrine, I have the answers that help anyone, in any context!

The Great Commission is not only "great" in scale, but also "great" in its importance!

Afterword

There is a non-contradictory model for God, faith, grace, and salvation. We have an understanding of why confession and repentance are the only way to remove guilt and acquire the value Jesus supplied. Every Christian should have this foundation; otherwise, they will be contradictory in their beliefs.

The Salvation Model relies heavily on righteousness. Grace is the best way to become righteous. The Reward Model relies heavily on justice. Rewards are the result of justice determining the amount of value. Both Models are guided by the universal principle of uniqueness.

We only have the beginnings of the Reward Model. Actually, the rest of sanctification is concerned with rewards. In order to add to this model, we would have to look more closely at our interactions with others. In order to complete our understanding of sanctification, the next book, Christian Living, looks more closely at how value is created and exchanged in our day-to-day lives.

We will see that both the Salvation and Reward Models are made complete and converge on love.

Joel Swokowski's Commentary

I see Book 1 as an archer pulling back a bow string. Book 2 releases that string and gives us all the information we need to be propelled forward into a magnificent walk with the Lord… or even better, a run to the mountain top!

Before we begin that book, I'd like to give additional insight into Book 1 which only occurred years after Modeling God was originally published.

The author was a huge fan of *Mere Christianity* by CS Lewis. In fact, the original preface began with him sharing that he believed the last great book explaining Christianity was *Mere Christianity*, and that 60+ years seemed to be as good a time as any to update it.

It all began with him reading *Mere Christianity* once a month for three and a half years (42 times!) and crossing out anything that seemed contradictory. He wanted to continue CS Lewis' legacy by writing a book that could be read a chapter a day each month (hence, the thirty chapters plus a preface).

Years later, he found that CS Lewis' last book was published shortly after he died. What did CS Lewis write in *The Discarded Image*? What was on Lewis' mind at the end of his life on earth?

Lewis believed we needed to create a Model (yes, he used a capital "M"!) for God that was large enough to cover all the effects of God, but simple and understandable. How did Lewis say we ought to create this Model? He gave several suggestions, which turned out to cover the principles of Modeletics!

Lenhart shared this publicly and even reached out to "experts" of CS Lewis. No one embraced this. How could people extol CS Lewis and intentionally ignore how he wished his legacy to continue? It was just another example of how people weren't ready for the truth presented in this book while confirming that the author wrote this book under the influence of the Holy Spirit.

It turns out that if people had truly continued Lewis' legacy, we wouldn't have seen the church in America go into decline. In 2009, the cover story for Newsweek magazine specifically tied the decline of the church in America to the "Godless Christianity" movement, which they say began in 1969 (six years after Lewis' death), when the church in America decided to go without an orthodox explanation for God's nature.

If you are interested in the details, go to this link: https://musicoflifechurch.com/know-god

The reality is that the book you are reading right now presents the only orthodox explanation for God and is the true continuation of CS Lewis' legacy.

Book 2

Christian Living, Modeling Life

PART ONE

The Journey

INSIDE

Chapter 1:   Profitability 206

Chapter 2:   Truth. 221

Chapter 3:   Weed Killer 233

Chapter 4:   Life 244

CHAPTER 1

Profitability

"ARE WE THERE yet?"

Almost everyone can remember what it was like to take a long car trip when they were a child. The focus was completely on the destination, while the journey itself was seen as a waste of time. It is only once we've become more mature that we realize the journey was also an opportunity for growth and enjoyment. In fact, the journey was critical. After all, it was possible to ruin our time at the destination or miss out on it entirely if we handled the journey poorly.

The first book in this series began with an analogy regarding an invitation to a party. The directions were given in an abstract fashion and were only useful once we got the address. Likewise, the first book focused on the destination (God's address and salvation). That is, we determined God's identity and the process for making it to His party. However, we didn't get a specific description of the party. For now, we referred to it as the ultimate relationship with God.

Before we can determine what the party is, we need to get to the party. We have been calling this process of getting closer to God "the journey." In this section, we will focus on answering the question, "What is the journey?" and on how we intentionally make and measure progress.

Progress

We have righteousness and justice to guide us; however, we aren't always going to be moving directly towards the party. We will have barriers in our path and may have to head away from the party in the short term in order to make the most progress toward the party in the long term. How can we tell which detours are temporary inconveniences and which detours will lead to an abandonment of our journey? It looks like we need an objective way to measure progress.

Since we are unique beings, we can't come up with a set of specific actions to measure this progress that will be the same for each individual. This is known as works. In the first book, we saw what Paul thought of doing the right thing for the wrong reason. While works result in rewards through justice, it does not result in salvation. Remember, the Salvation and Reward Models are separate. We will need a universal principle based on righteousness, justice, and uniqueness that measures our progress. Furthermore, in order for God to remain holy, there must be a universal measure based on our uniqueness that God can objectively use to determine who will get saved and continue the journey.

We saw with justice that determining a model applying to all conditions requires us to identify a universal principle. The key to identifying this principle is to continue stepping back and taking a big-picture view. Recall that it looks to most people like justice doesn't exist because we see good people die young. Yet, we realize that justice does exist when we step back and see that their death isn't the end of receiving value.

The biggest mistake people make in modeling is to stop short and declare something as a universal principle, and require people to obey it. This action ends up denying the universal principle of uniqueness. We've seen that righteousness and justice are universal principles that apply to all of us. In the same way, we find that profitability is a universal principle and it is used to measure progress.

We saw in the chapter on uniqueness that Jesus described the ability to enter the kingdom of heaven with the parable of the talents. Three people were given different amounts. After a time, God assessed what they did. The two who created more were let into heaven. The one who created nothing was sent to hell. This person would have made it to heaven if he had simply generated interest (Matthew 25:27). That is, all the people were judged regarding their ability to generate a profit. In fact, Jesus called evil servants "unprofitable" (Matthew 25:30).

Profitability

The importance of profitability is obvious to every person in the world. For example, profitability is a universal principle that has been used to measure the progress of mankind. All of us consume, whether it is food, energy, etc. In order for any endeavor to survive over the long term, it must produce more than it consumes. Think about it—our country couldn't survive on what was produced 200 years ago. In order to create more, we must use the principle of profitability.

Profitability is the difference in value between what an object is worth and what the object cost to create: Profitability = Worth - Cost of creation

Notice that worth is not simply the sum of the raw materials and the labor (time) needed to make the object. Since the time needed to make the object has a value, this would result in breaking even. Profitability is everything gained beyond what it costs, and that doesn't necessarily have to be money. Also, keeping with our focus on being intentional, profitability doesn't just happen. It must be a goal in order for it to be achieved.

Profitability is completely dependent on uniqueness. If everyone valued everything exactly the same, there would be no way for people to become profitable. Profitability is the effect resulting from people exchanging value for value, but the individuals have to value things at different amounts in order for everyone to be profitable.

For instance, let's say someone has an object he or she values at $5, while you value it at $15. If you buy it for $10, both parties actually become profitable by $5. However, if both parties valued it the same ($10), then they couldn't be profitable.

The Creator built profitability into this world when unique beings were created. Trying to get people to value things the same goes against uniqueness and profitability. We need people to value things differently. Profitability is at the heart of the difference between two distinct philosophies about how life is sustained over the long term.

Two Philosophies

The first philosophy states that resources are finite, and humans consume resources. Therefore, survival is dependent on minimizing consumption. The central belief of this philosophy is to deny the self. Go without. The only time resources are consumed is to avoid death. Everything about the society tries to minimize progress. This necessitates the need to isolate the society from the rest of the world and go into maintenance mode.

This could be done in the past, but not today. The world has matured towards interdependence. In fact, today these societies have shown progress in technology and defense. It should be obvious to these societies that they are contradicting their core beliefs. As we've learned, contradictions cannot exist in reality. Therefore, this is a flawed philosophy because they can't follow it completely and survive. Besides, what is the end result of this philosophy? Death. Eventually, all the resources will be consumed and everyone will die. This philosophy just slows down the inevitable and focuses on survival. I call this the "Survival" philosophy.

The second philosophy states that resources are infinite if they are created (produced) by the humans who consume them. The ultimate goal is for society as a whole to produce more than they consume. How is this done? In the short term, resources are consumed not only for survival but also to generate resources of more value. We called this profitability.

In the long term, effort should be spent to develop more efficient ways to use resources so that fewer resources are used, like in the first philosophy. Notice, this is second because efficiencies can only be generated once the process is well understood, and that takes time. For now, let's call this the "Profitability" philosophy.

This Profitability philosophy is much better than the Survival philosophy at ensuring the minimum quantity of resources is used. Why isn't the Profitability philosophy obvious? There are a lot of reasons. The biggest one is that people look at the short term. This Profitability philosophy is risky and could result in a quicker death if it isn't done well. If people chase after short-term profitability, it can result in a long-term decline in profitability, and a quicker death.

The Survival philosophy is appealing because it is safer and more comfortable for people to prolong an inevitable death. The Profitability philosophy requires thought and intentional effort to create new methods. As we've seen, growth is not comfortable.

Survival Philosophy

1. Resources are finite

2. Humans consume resources

3. Minimize consumption to slow down the inevitable: death

4. Long Term: Become more efficient

Profitability Philosophy

1. Resources are infinite if humans produce

2. Humans consume resources

3. Short Term: Use resources to generate resources of more value

Bible Examples

When I mentioned the need for a universal measure to determine who continues the journey after physical death, you may have wanted to call this "good." You could have said, "Good people go to heaven; bad people don't." My response would have been to ask you how you would objectively measure "good."

Actually, some form of the word "profit" is used throughout the entire Bible as the universal measure of good. We are all familiar with profit as it relates to money, but the word in the Bible applies to that and more. For example, Jesus makes profitability (which He called "bearing fruit") the universal measure to determine whether people are going to heaven or hell.

> "I am the true vine, and my Father is the husbandman… Every branch in me that beareth not fruit he taketh away: and every branch that beareth fruit, he purgeth it, that it may bring forth more fruit… If a man abide not in me, he is cast forth as a branch, and is withered; and men gather them, and cast them into the fire, and they are burned… Herein is my Father glorified, that ye bear much fruit; so shall ye be my disciples" (John 15:1-8).

Next to the kingdom of God, Jesus spoke about money more than any other topic. Jesus used examples using money to illustrate the importance of profitability in an undeniable and immediate way. Here's an excerpt from the parable of the talents that was covered in the chapter on uniqueness from Book 1:

> "For unto every one that hath shall be given, and he shall have abundance: but from him that hath not shall be taken away even that which he hath. And cast ye the unprofitable servant into outer darkness: there shall be weeping and gnashing of teeth" (Matthew 25:29-30).

We see here that being unprofitable is bad. Another term for unprofitable is "vain." There are several passages in the Bible that speak poorly of people doing things in vain; that is, without a profit. Notice, the previously mentioned passages relate to the Salvation Model. Since God is the ultimate source of the value created, we can't claim any credit for the result.

Another place the Bible speaks of profitability is when it talks about the body. It sees people as different parts of the body. The body as a whole is profitable because of all of its parts. So, everyone can contribute a value. We've seen the same perspective in the chapter on uniqueness.

Ultimately, the cause of the value of the individual is their uniqueness, which was given by God:

> "Nay, much more those members of the body, which seem to be more feeble, are necessary: And those members of the body, which we think to be less honourable, upon these we bestow more abundant honour; and our uncomely parts have more abundant comeliness. For our comely parts have no need: but God hath tempered the body together, having given more abundant honour to that part which lacked: That there should be no schism in the body; but that the members should have the same care one for another. And whether one member suffer, all the members suffer with it; or one member be honoured, all the members rejoice with it. Now ye are the body of Christ, and members in particular. And God hath set some in the church, first apostles, secondarily prophets, thirdly teachers, after that miracles, then gifts of healings, helps, governments, diversities of tongues" (1 Corinthians 12:22–28).

Notice the parts that lack are given more honor and are actually the most important to the body. Why? The profitability of the body as a whole is dependent on the amount it consumes and the amount it produces. There is a limit to the amount some parts of the body can produce. In fact, the passage says the parts that produce more than they consume "have no need." However, there is no limit to the amount some parts of the body can consume. In reality, the parts of the body that lack have a greater impact on the profitability of the body as a whole.

Likewise, in our society, the members that consume the most without producing are actually the most important and deserve the most attention because they have the greatest impact on the profitability of us as a whole. This passage says we are all affected by the suffering of others and their lack of profitability affects us all because we are all connected.

Notice that the Profitability philosophy accounts for interactions between people. The Survival philosophy depends on people remaining isolated. The first book looked at interactions as a zero-sum game through justice. In order for one person to gain, someone had to lose. In this book, we will see that uniqueness allows for people to interact in a fashion where everyone gains through profitability.

We saw in the parable of the talents that profitability is the result of two things:

1. People valuing things differently (uniqueness).
2. People interacting with each other.

The unprofitable servant isolated himself by burying the talent. Jesus said the servant would have gone to heaven if he had interacted with just one person: the banker.

We can tell if we are making progress on our journey by examining and measuring the long-term profitability of our actions. Since profitability is an intentional action, it would be best for everyone to understand how to be profitable. Profitability is based on uniqueness and is an act of creation. We've looked at uniqueness. In the next chapter, we will look at what it takes to create.

- Describe an interaction that was profitable.
- What did each person gain?
- What areas did each person value differently?
- What resources are you consuming without getting a return?

- What resources are you investing and what are you getting in return that is of greater value?
- How would you explain to someone the importance to all of us of helping those who are unprofitable?

Joel Swokowski's *Commentary*

In order to better understand the concepts taught in this book, allow me to explain a universal tool that will ensure profitability and even guide us into truth. This tool will be especially helpful when we get to the next chapter: Truth. The tool: Conjunctives.

Conjunctive: A logical statement in which both halves must be true for the statement to be true. In logic, this is known as the "and" statement.

Let's look at some examples of conjunctives in the Bible.

"Let all things be done decently and in order" (1 Corinthians 14:40).

1. Let all things be done
2. Do things decently and in order

Notice that there isn't one specific way to fulfill this verse to ensure profitability. There is "wiggle room" that can be considered according to the uniqueness of the individual(s) and the circumstances; however, the wiggle room is bound on one side by part A **and** on the other side by part B.

Actually, if there are two people allowing their flesh to go in different directions, they use the other to feel justified. For example, one pastor can state he **only** wants to do the things of the Spirit that can be done in order. So, the pastor allows prayer for healing, but he won't allow anyone to prophesy. His excuse is that he doesn't think people can handle this without being destructive…and he knows God wouldn't want the church to be destructive. This pastor feels like he has followed the Word of God even though he is only doing **half** of the conjunctive. He feels those pastors who are allowing everything are more wrong than him and destructive.

A second pastor can allow **everything** to be done and doesn't correct people when they are out of order. His thought is that he is doing the Word of God because he is doing **all** things…and people are going to make mistakes so God wouldn't expect everyone to do everything in order all the time. More to the point, this pastor would say he is clearly following God's Word more than the first pastor because he is encouraging **all** things to be done. Even if he admits he isn't doing the whole Word of God, he's better off than the first pastor.

Both pastors are not doing the Word of God. Worse, they are distracting the rest of us by arguing over two perspectives as if one of them must be the truth, making all of us unprofitable.

This verse was intended to encourage people to let the Holy Spirit flow through them **and** rely on the pastor to confront people to keep things in order. Neither pastor is doing both halves of the conjunctive, so neither pastor is doing the Word of God…and if they say they are, then they are false teachers.

When we look more closely at a conjunctive as it relates to the topic of this chapter, we see that one half of the conjunctive gives the benefit and/or freedom. This is the half of the conjunctive that we ought to drive as much of our resources into as possible. There's no limit that is too much! The more we do, the more profitability we will achieve. For instance, it was "all things" in the previous example.

The other half of the conjunctive gives one limitation (which can also be a punishment). This is the half of the conjunctive that is like a check in a box. There's a right amount that we need to achieve. Any more than that limits our profitability, with too much causing us to be unprofitable. Not enough and it doesn't reach the limitation allowing us to be in the conjunctive ensuring profitability. In our previous example, it was "decent/in order".

"All things" = do as much as possible
"decent/in order" = check the box

Notice, the freedom half of the conjunctive can be seen as an animal (fleshly) thought process when considered by itself (without overlap). The limitation half can be seen as a human (logical) thought process when considered by itself. Here is how this is represented with a diagram:

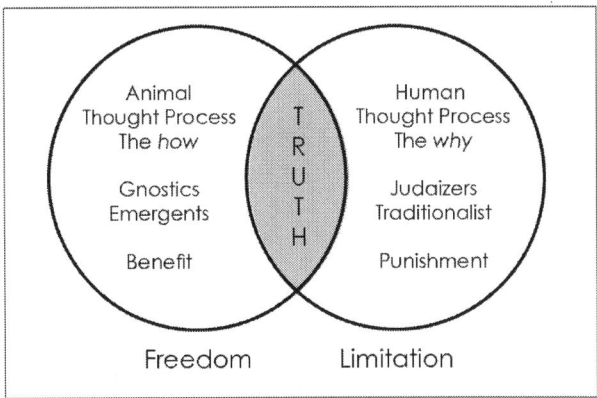

CONJUNCTIVE INTRODUCTION

Another way to look at the conjunctive is the **limitation** that results in something being created. It can be seen as the right **why** for the benefit. The **benefit** portion of the conjunctive can be seen as the right **how** for the limitation.

We can use this diagram for any truth, any doctrine! Check this out…

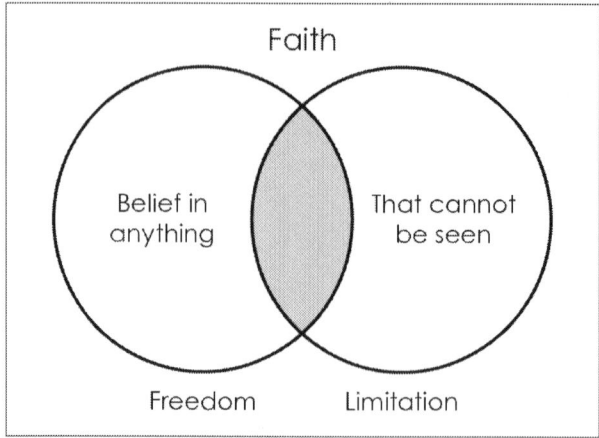
CONJUNCTIVE ILLUSTRATION FOR FAITH

Notice, a "belief in anything" is the freedom while "that I cannot see" is the limitation. A belief I **can** see would not be faith. Something I cannot see that I **don't** believe would not be faith.

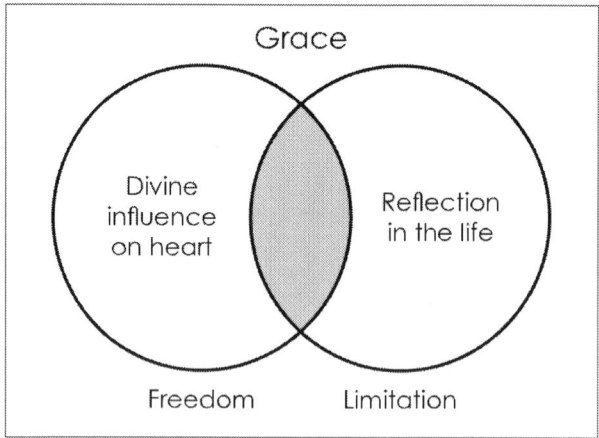
CONJUNCTIVE ILLUSTRATION FOR GRACE

The "divine influence upon the heart" is the freedom, while "reflected in the life" is the limitation. Any divine influence that is not reflected in my life would **not** be grace. Also, some behavior reflected in my life that is not caused by the divine influence is **not** grace.

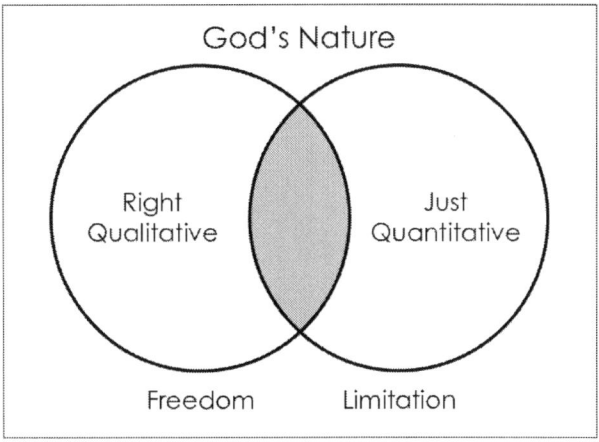

CONJUNCTIVE ILLUSTRATION FOR GOD'S NATURE

Righteous is the qualitative and therefore the freedom. Just is the quantitative and is, therefore, the limitation. There are varying degrees of justness, so as long as the act doesn't violate justice, it will create. A "god" that is only right and not just is no god at all. That god wouldn't always be profitable. The same could be said for a "god" that is just but not right. That god couldn't create.

Notice that the **truth** is where both halves overlap. The portion of the **freedom circle** that isn't truth would represent people like today's emergent and the Gnostics (during Biblical times). These people deny limitations and can't logically explain what they believe without contradiction. This is the definition of the **animal/fleshly thought process**, which we know does not create.

The portion of the **limitation circle** that isn't truth would represent people like today's traditionalists, and the Judaizers (during Biblical times). This is the definition of the **human/logical thought process**. People who believe it is right to limit others instead of doing the truth because they are preventing people from being destructive in an animal/fleshly thought process.

The Profitability Philosophy is a Conjunctive: eat enough to survive and plant/invest as much as possible.

"Plant/Invest" is the Freedom = do as much as possible
"Eat enough" is the Limitation = check the box (survive)

CHAPTER 2

Truth

DUALISM IS THE belief that there are two equal powers at war. One power is what we call "good" and one is what we call "bad." However, the bad power cannot survive without having qualities that we call "good," such as intelligence or strength. So the bad power has qualities we would call "good," but he just doesn't use them to "do good."

Dualism sounds logical until someone asks, "Where would the bad power get these good qualities?" They must have gotten them from the good power, so the good power must have created the bad power. Therefore, you can't have a good power and an evil power that are equal. In fact, the evil power depends on the good power.

There are a couple of implications that should be addressed. First of all, we know the good power has the ability to create because the good power created the bad power. Secondly, what does it mean to "do good"? We saw the bad power doesn't use good abilities to "do good"—otherwise, it would not be the bad power. What we mean by "do good" is to "create." The bad power doesn't create; it only destroys.

In the Bible, we see that God (the good power) is referred to as the Creator. We also see God created Satan (the bad power). Jesus says that Satan is a thief that looks to steal, kill, and destroy (John 10:10). To put it simply, good creates and evil destroys. You've probably heard people say you can't have good without evil. That is incorrect. The good power creates and can exist without the evil power—isn't the lack of an evil power one of the benefits of heaven? However, you can't have evil without good, because you must first have something created before it can be destroyed. Evil's existence depends on good.

Finally, some people think the ability to destroy an object makes them equal to the person who created it. One form of this belief is seen when people focus on the shortcomings of others rather than give a solution on how to make the person better. They tear others down in hopes of making themselves look equal or better. This also is incorrect. For instance, anybody can destroy a work of art. It takes a unique individual to create the work of art.

Profitability is the creation of value that occurs when something is exchanged between two entities (interaction) that value the thing differently (uniqueness). Uniqueness is vital because we need people who value things differently. Creation and profitability are interdependent. Profitability is the creation of value and creating is the act of making a profitable thing— something that didn't exist before that has greater value than its original components.

Making something that didn't exist before out of more valuable components is actually an act of consumption in the big picture, not production. Let's look at another quality that distinguishes good from evil.

Truth

Pilate asked Jesus, "What is truth?" (John 18:38). Some people say, "Truth is reality." Actually, reality applies to things that exist. However, there are things that are real, that exist, and that aren't truth. Things can be real and not be true. Likewise, things can be true and not real. So truth must be something more than reality.

Another definition people use is "truth is facts." Truth is certainly factual. However, are facts the same as truth? Satan spoke facts to Jesus when he quoted the Bible, but we saw that Jesus says there is no truth in him at all (John 8:44). So truth must be something more than simply "facts."

Truth has to be different in a way that goes beyond the actual words or concepts stated because we know that the Bible is truth, yet it wasn't truth when Satan quoted it. The only thing left is the purpose of the words. Truth must depend on the intent of the user of the facts.

What was Satan's intent? It was to take Jesus off His goal. It was to make Jesus unprofitable; that's always Satan's goal. Satan quoted facts to Jesus. Facts are good. However, Satan wanted to use facts in order to destroy. He wanted to use a good object for an evil purpose. Once again, we get a picture of the problem with dualism. Evil takes good attributes and uses them for the wrong purpose—to destroy. Jesus has truth because He uses facts to create. God's Word is truth because when God speaks it, it results in creation.

How can we define truth in a non-contradictory fashion? Truth is "facts that create in the long term." In order to create something of value, you must use facts. You can't create anything of value by using things that aren't factual. For instance, you couldn't solve a math word problem by using calculations or arithmetic operations that are incorrect.

When someone quotes a fact, it is only truth if the fact creates something of value. For example, the Bible is fact. However, the Bible is only truth when it is used to create. Simple facts can oppose other facts. Simple facts can even contradict each other. Recall that we remove contradictions by taking a big-picture view beyond the realm of the simple facts. However, truth cannot oppose itself. Truth is non-contradictory.

Creation can only occur with truth. We can say creation is an effect of truth. Since creation and profitability are interconnected, profitability is an effect of truth.

Implications

There are other implications and applications of truth. First of all, notice that it is possible for short-term destruction to be a good thing when we account for long-term profitability. For instance, a person may need to tear down a condemned building in order to build something of use. The initial destruction, while it results in short-term chaos, is seen as good because it is part of a bigger objective.

Likewise, people say, "The truth will set you free—but first it is going to make you mad." Truth makes us uncomfortable in the short term when it tears down something we've built. However, it is just the first step in a bigger process of growth and, eventually, profitability.

We are now able to better explain the relationship between knowledge, understanding, and wisdom. We said that knowledge is facts. We also said understanding resulted in knowledge. Another way of saying this is to say that understanding creates knowledge.

Notice, Satan quoted facts (knowledge) to Jesus. In fact, Satan can only quote knowledge. Satan has no understanding. If he did, he would be able to create. We saw in the first book that knowledge leads to pride. The Bible says in Ezekiel 28 that Satan fell because of pride.

In fact, this is a good place to review "faith." Notice, Satan does not have faith. He does not have profitable actions (wisdom). Why? He has no understanding. However, he has experience and knowledge.

We have the ability to understand. We have this ability to create because of God. If we believe it comes from ourselves, we are in pride. Notice, Satan actually depends on God (through us) for understanding. People who say Satan has faith, is wise, or tells half-truths are proving their lack of understanding with their presentation of these contradictions.

A familiar place to look for the application of truth is with stories. Some fiction writers believe the only way to write about complex truths is through fantasy (e.g., *Lord of the Rings*, *Harry Potter*, etc.). In the writers' opinion, reality is too confining. In fact, reality can be distracting. Rather than grasp the universal truth, the audience can get caught up in the details. This is especially true when the story is too close to their reality. This familiarity is actually a hindrance to the individual's ability to see the big picture.

When a story is fantasy, the audience lets go of its frame of reference and looks at the message with new eyes. There is nothing wrong with fictional fantasy stories in and of themselves. In fact, Jesus told parables to get the same effect. In some of the stories, He posed a situation the hearer thought was unrelated to their experience.

Sometimes, Jesus would even end the story with a question. This process allowed the listener to see the big picture and assess truth. Once the listener responded, Jesus would sometimes show the listener how their answer was the opposite of their actions. While the moment may have been uncomfortable, it eventually led to long-term growth.

Truth can be understood apart from reality. For instance, people can receive truth through dreams or by using their imagination. Let's face it; even stories that are based on reality aren't real. For instance, with movies, TV shows, and plays, do we really think what is happening is "really" happening? We know actors portray the characters. So, even some suspension of disbelief is needed in order to receive truth. Because this suspension of disbelief is deliberate, the hearer has to be intentional in order to receive truth.

Finally, equating "facts" and "truth" can lead to numerous destructive situations. For instance, a husband could tell his wife all the things she doesn't do as well as other women. When she becomes upset, he could say, "I'm just being truthful." Actually, the husband is being factual. It is only truthful if it creates a value in the long term. The Bible doesn't require us to speak only facts; it demands more of us. It tells us to speak truth.

Art

We've been using examples from art. What is a non-contradictory definition of art? We have seen that an artist creates, that is why they are called "creative." What do they create?

Art is the individual's unique expression of truth. People express truth in different ways because they are unique. Remember, everything of value is fueled by truth. If it is art, then the truth within it fuels it. Ultimately, it needs no energy from the outside to sustain it and will outlast its creator because the audience chooses to internalize the truth. In fact, the same truth can be related to in different ways because of the uniqueness of the audience members.

What is the opposite? What do we call it when someone expresses what he or she believes as truth but it requires constant energy from the outside to sustain it? We call this propaganda. Notice, when the person who created the propaganda is removed, the energy needed to keep it going is gone and the movement dies.

If you study artists long enough, you will find they are very intentional in their methods. This confuses people because it is usually not a method shared by others. However, that is a measure of its uniqueness, not truthfulness. Unfortunately, the uniqueness of the method can cause people to reject truth because of the source. If you understand that truth is non-contradictory, you will not discriminate against sources of truth.

Good Model
1. Profitable in the long term
2. May destroy in the short term
3. Uses truth
4. Can exist without evil
5. Non-contradictory

Evil Model
1. Unprofitable in the long term
2. May appear to create in the short term
3. Uses facts only
4. Can't exist without good
5. Contradictory

Summary

Profitability is God's universal measure for determining whether people are allowed to continue the journey once they die; however, we've seen two models whose result is creation of value. The Reward Model accounts for everything we do, regardless of the reason. According to it, we can gain by others destroying, even if we initiate the destruction (Figure 7, Panel 4).

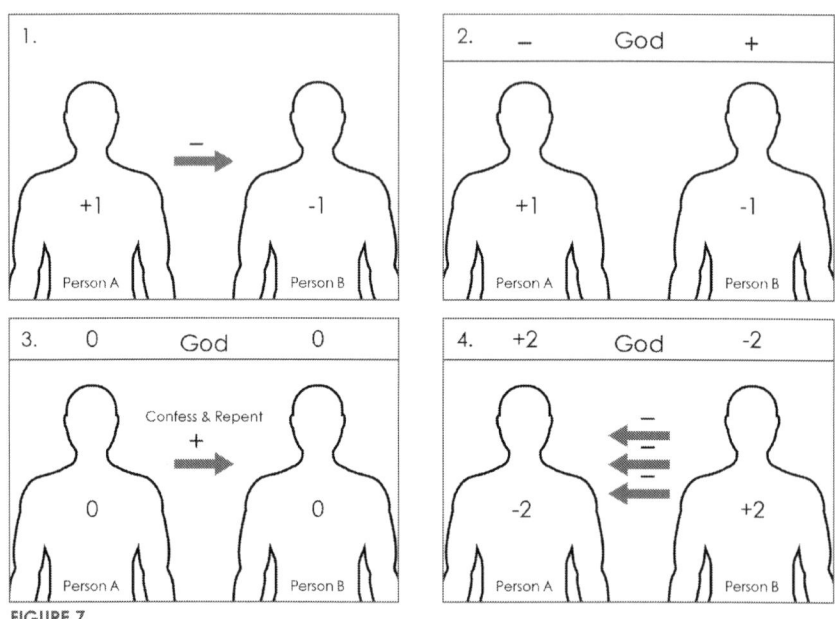

FIGURE 7

Sometimes the Reward Model doesn't result in long-term profitability because for us to gain, someone has to lose. Justice's focus is on equaling everything out, not profitability. The Salvation Model started with us getting a value from God and ended with us choosing to let it come out in our actions. The action always results in long-term profitability for everyone.

Long-term profitability is achieved by using truth. Since truth is facts that create, profitability can be measured by how much an individual is creating during their journey. It is an objective measure. If the individual is bearing fruit (creating), he is progressing on the journey. If the individual is consuming resources and producing something of less value (destroying), he is showing God he does not want to continue the journey when he dies.

Truth is a cause the individual uses to be profitable. Our eyes should be on truth regardless of the source, because it always leads to long-term profitability. Consequently, truth is the individual's best guide for the journey. Ultimately, truth can only come from God. All truth is God's truth. Therefore, salvation can only originate from God and we can't claim any credit. There is nothing we can do to make God value us more or less.

Remember, while God initiates reward, it seems most of the time rewards are initiated apart from God. This is just one reason why we can't equate blessings with absolute proof we are in the will of God, and we can't equate trials with absolute proof we are outside of the will of God. Blessings and rewards can be initiated apart from God and they have no bearing on how God values us.

The first book showed us we could be profitable through justice when others take a value from us. In order for someone to gain in these interactions, someone had to lose. This book has shown us a "loophole." Now we see a way for everyone to gain from an interaction. Value can be created if unique individuals interact in truth.

Finally, we see why a god can't be a Creator unless He is righteous and just. In order to create and be the Creator, He would have to be right. The Creator would have to make sure what was created was of more value than what He started with; otherwise, the process would be destructive. The quantifiable principle of justice would be necessary to determine if creation is the result. Let's look at more implications of truth in the next chapter.

- Describe how money, sex, and medicine can be "good."
- Describe how these "good" things can become "bad."
- Think about the last argument you had.
- What did you say in that argument that was factual?
- What did you say in that argument that was truth?
- What truth can you find in Buddhism, Islam, or some belief other than Christianity?

Joel Swokowski's *Commentary*

"Good creates, evil destroys." This chapter does a great job explaining how we may need, and often need, short-term destruction in order to have long-term good or long-term profitability. This explains why the book of Isaiah states the following: "I form the light, and create darkness: I make peace, and create evil: I the Lord do all these things" (Isaiah 45:7).

Truth is defined as "facts that create in the long term." It's also clear that truth must depend on the intent of the user of the facts that they have. Notice, this is a conjunctive. Facts are the freedom and creating in the long term is the limitation.

We've already seen the difference between a right *what*, a right *why*, and a right *how*:

What = fact, knowledge
Why = cause, understanding
How = doctrine, wisdom

If we take this model and apply it to truth, what we see is:

> Truth is "a right *what* with a right *why* and a right *how*."
> Truth is "facts (right *what*) that create in the long term (right *why/how*).

When we look at this from the perspective of a conjunctive:

> The "freedom" is the right *what*.
> The "limitation" is the right *how* (because the right *why* would have to be known).

A right *why* by itself without the *how* is the limitation half by itself. The *how* is not what the person did or how the person stated it, that would be *what* they did or *what* they stated. The *how* is the understanding of *how* the concept works; the manner in which the concept was applied; the process; the method; the model.

This also gives us an explanation as to truth depending on the intent of the user. The *why* and *how* would be the intent of the user. *Why* are you stating this fact with respect to *how* you think it works?

Another benefit of this model is it helps us understand deception. Deception would be "a right *what* with a wrong or no *why* and *how*." The reason people get deceived is that they only look at the facts. Understanding the causes and knowing doctrine (wisdom) is the way to avoid being deceived. Without realizing it, deception is what people are actually describing when they say something is a "half-truth."

A lie would be a person intentionally giving a wrong *what*. A lie is evil because it intentionally prevents creation and can only result in destruction.

We also recognize that all truth belongs to God. He is the Creator, His Spirit is the Spirit of Truth. The vessel in which truth is delivered will be unique, so I ought to at least be open to truth regardless of the source with the faith that if it is truth being spoken, the original source is always God. This reminds me of the question about a person who said God spoke to them from a tree. Our first response ought to be, "What did He say?" Truth is the measure for wisdom which Lenhart spoke about in chapter 6 of the first book.

Finally, let's look at conjunctives again. A conjunctive is a measure for truth. All truth is a conjunctive, and we can even see that the definition of truth is a conjunctive itself: Facts (freedom) that create in the long term (limitation).

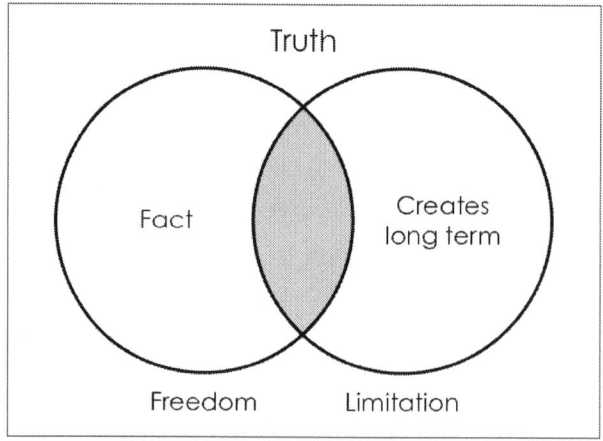

CONJUNCTIVE ILLUSTRATION FOR TRUTH

Every non-contradictory definition will be stated in the form of a conjunctive.

CHAPTER 3

Weed Killer

WHEN I BUILD mathematical models, I pay close attention to whether the model is getting tighter or looser as more data is added. Is the model able to handle all situations and predict outcomes, or do I need to add a lot of special cases to handle the contradictions?

In fact, this is the easy way to blow up a bad model: feed it and make it try to hold up under the weight of all the data. When there isn't enough truth to hold up all the contradictions, the model fails.

I remember this by thinking of weed killers. Modern-day weed killers don't work by "killing" growth. It is just the opposite. Today, the most successful weed killers accelerate the growth of one aspect of the plant faster than the other aspects can keep up. It is a fertilizer! Long-term growth depends on several aspects working in harmony. The weed essentially dies because one aspect grew too fast.

Resisting Evil

In the same way, evil is full of contradictions and will eventually choke itself. Evil is unsound and will disintegrate. It can't hold up in the long term because it doesn't create. Evil needs someone to create for it in order to have something to destroy.

The quickest way to expose and destroy evil is to speed it up. When you resist evil, you are actually making it stronger because you are giving it your energy by being evil. Don't fight it. That is, don't take actions that don't create. When you fight evil, it requires you to be evil, because you are not creating.

The Bible says, "Be not overcome of evil, but overcome evil with good" (Romans 12:21). This means we should help evil become profitable! Since we know evil's goal is destruction, anything we do to help it will only end in its destruction.

If we are mistaken about evil, the good part will end up being profitable. Notice justice is determining our actions. Regardless of whether we are right or wrong about identifying evil, we won't receive a penalty because we were helpful. In fact, if we are wrong, then we still get a reward because profitability will result.

There is another implication for how we are to interact with evil. Evil wants to distract us from our goal. Evil doesn't want us to be profitable. Since evil doesn't create, it needs someone else to do that. Evil depends on our consent in order to function. If you do not give it your consent, it cannot use you.

Ultimately, you are responsible for the evil that affects the part of you that continues the journey after you die, because you are the only one who can let it affect the eternal you. Nothing else can. Paul said in his contrastive fashion, "For I am persuaded, that neither death, nor life, nor angels, nor principalities, nor powers, nor things present, nor things to come, [N]or height, nor depth, nor any other creature, shall be able to separate us from the love of God, which is in Christ Jesus our Lord" (Romans 8:38-39).

So if we are separated, what's left to be the cause other than ourselves?

When we encounter evil, we are not supposed to give it our consent to exist by following it in our actions. We are not supposed to energize it by fighting it, that is, take evil actions against it. We are supposed to deal with it intellectually; agree with it in order that we can be good to it by facilitating its growth intellectually (understanding). If it is good, it will flourish. If it is evil, it will dissolve from the acceleration of growth, just like a weed.

Remember, our goal is to get to the party by proving to God that our will is to continue the journey after we die. Make no mistake about this—evil will affect our physical selves. If we let evil affect our spiritual selves by making us unprofitable, we are telling God that our will is not to continue the journey as a spiritual being after we physically die.

Model Building

Basically, we are trying to identify a model to explain how to get closer to (grow in our relationship with) God. In order to determine this model, I am looking for non-contradictory facts that create. I'm looking for truth. Some of that truth will come from my resolution of contradictions. Some of it is going to come from other people stressing my model.

If my goal is to be right and not have to change my model, then I will try to dismiss others who have something to say about this model. I will look comparatively rather than contrastively. If my goal is to identify a tighter model, there is the distinct possibility that truth will come from someone outside me, unless I'm perfect.

Let me be straight right now; I'm far from perfect.

However, some of the people who come to me won't be contributing truth. How do I tell the difference between those pursuing truth and those pursuing their point? How can you have profitable interactions with people whose beliefs differ from yours?

Disagreement

We have mentioned there will be hindrances on our journey. As much as we discussed Satan in the last chapter, one of the biggest hindrances on our journey is other people. Think about it—there are other people on a journey. They are making decisions on how to get to the party. When they see you going in a different direction, someone is going to feel like they are wrong. Someone is going to try to convince the other person that they are right. How should we deal with people who disagree with us?

It is the same perspective as model building. First, focus on pursuing the truth. Creation is achieved through truth. People who pursue growth focus on pursuing truth. People who pursue comfort focus on proving their point. Next, we need to find out whom we're dealing with. People who are pursuing truth will help us grow or we will help them grow. Let's look at the people proving their point.

In order for the comfort-driven to convince themselves and others of their point, they must present truth in an equal or greater quantity as their flawed point. For example, we see this when politicians state a truth and then try to prove a separate untrue point. If their untrue point outweighs the amount of truth in their presentation, we don't believe their position. When this deception does work, it is because people make the mistake of believing everything someone says when he or she recognizes any truth in the presentation.

Deconfirmatory logic tells us to try to disprove everything. Whatever is proven wrong should be discarded. Whatever survives the scrutiny should be retained. Paul realized the importance of handling this deception and wrote about this twice:

> "Study to shew thyself approved unto God, a workman that needeth not to be ashamed, rightly dividing the word of truth" (2 Timothy 2:15).

> "Prove all things; hold fast that which is good" (1 Thessalonians 5:21).

In the first passage, Paul is saying the measure of our ability depends on how well we are able to separate out truth from among everything else we receive. In the second passage, Paul is urging us to be contrastive in order to prove all things we receive and only hold on to what is good, that is, what creates. Paul clearly understood the way to intentionally make progress on our journey.

Debate

Paul's advice is particularly useful when we deal with isolated statements and/or comprehensive philosophy. We can parse the concept into sections of truth and untruth. We can choose what to believe and what to discard. We can keep this confrontational mentality inside ourselves. However, when it comes to debating with people, this process will only antagonize others. What does Jesus say we should do when we disagree with people?

"Agree with thine adversary quickly" (Matthew 5:25).

"But I say unto you, [T]hat ye resist not evil" (Matthew 5:39).

Notice, this technique actually requires that you agree with the person! You should take on their beliefs. This is the good way to debate because it creates. It is also the quickest way to prove someone wrong. Why? If your goal is to make progress, resisting them is going to cause you both to go nowhere and be unprofitable. You will be at a standstill. It would be more profitable to go down their road and try to find a situation where their belief is true.

For example, people brought a woman to Jesus they claimed should be stoned according to the law because she was an adulterer (John 8:3–9). They asked Jesus what to do. He could have resisted them and told them not to do it. That would have resulted in an argument about the law. He could have run around trying to prevent the crowd from stoning her by knocking the rocks out of their hands. That would have gotten them riled up to stone her.

Instead, Jesus took on their beliefs and tried to come up with a way to stone her! He said the one without sin should cast the first stone. This dispersed the crowd more effectively than any other method. Jesus didn't resist evil; He agreed with it and helped it reach its end more quickly. Unfortunately, some people today believe that just considering another belief is a sin in itself.

Win, Win

Try this the next time you're having a debate—take the other person's side and try to apply his or her point to other areas (big picture) or into the future (long term). Taking the other person's side immediately lowers his or her defenses. You are also keeping this deconfirmatory process internal by trying to prove your own beliefs wrong by accepting the other beliefs.

If the other point is true, it will hold up and may even explain situations in other areas, in effect creating knowledge. This would result in that model becoming tighter. If you are interested in growth, you will have achieved it because you will have learned something. You win if they are right! People who won't consider another belief are pursuing comfort over growth. They will resist understanding and stick to quoting facts without explanation.

If their point doesn't hold up, what you've done is made the amount of flawed perspective greater than the truth it was presented with. The presenter now has three options: add more truth, cut back the model, or admit they were wrong. Adding more truth tends to result in a looser model filled with special cases. Usually, people cut back the model by saying, "I never meant for it to apply to that situation." Again, you win! Ultimately, there is no reason to argue with anyone. Personally, I don't have long arguments anymore.

I believe God is very real and wants to help us grow; however, God is prohibited from helping certain people because they are under mercy. I also believe people realize this subconsciously and try to stay under mercy. The only way they can do this is to not express their will. Get them to state their will. The following technique becomes particularly useful when you try to correct people. Let's look at an example.

If someone promised to do something for me and didn't do it, the typical way to handle this is to tell the person what they did was wrong. We've seen in the first book that the person already knows that what they did was wrong and they are experiencing guilt, which leads to pain. When I confront them, I am giving them a reason to unload their pain onto me and they will not change their behavior because I'm "worse" than them. This is the short-term justice strategy.

Taking everything we've learned from the weed killer strategy, the more effective way would be to ask them three questions. I'm increasing their understanding. First, I ask them if they promised they would do the specific act. Second, I ask them if they did the act. Third, I ask if they think it is okay with God that they didn't do the act.

If you try this, your first goal is to get a yes or no for each question. In fact, tell the person the Bible says to let your yes be yes and your no be no (Matthew 5:37; James 5:12). You may quickly find out that they don't want to give a yes or a no.

If they want to give an excuse, tell them you will take an explanation after they say yes or no. If they say you are going to respond a certain way if they give one of those answers, assure them you won't respond and will walk away immediately after getting all three answers. We are not supposed to strike back because it limits God's ability to get involved. It is to your benefit to walk away, because it is the same as forgiving the person; they will have stated their will, and you are walking away, in effect, putting them in God's hands.

You aren't due a penalty because you haven't done anything unjust; you've just taken them out from under mercy. Justice says God will act on them if they are in the wrong because they have expressed their will that they believe something that is wrong. God is then required to help them grow. If they are in the right, there is no penalty either way.

Disagreement Model
1. Pursue truth
2. Determine if the person is pursuing truth or their point
3. Agree with the person
4. Take on their beliefs
5. Apply the belief to other areas (big picture)
6. Apply the belief into the future (long term)
7. Have them state their will and then you walk away

- In your last disagreement were you proving your point or pursuing truth?
- How would you handle that disagreement differently today?
- Think of a topic you often find yourself arguing about.
- What part of the opposition's argument is truthful?
- What part of your argument doesn't hold up in the big picture and/or long term?
- When is an argument no longer profitable?

Joel Swokowski's *Commentary*

"Weed killer" is an application of Contrastive Thinking. It is the quickest way to find the flaws or weak points in the other, even the opposite, perspective. You essentially see yourself as **having** to defend the other position. You will see the strengths, but the weaknesses will unsettle you and you will realize the quickest way to expose the weakness.

The best way to become more right is to find where I'm wrong. Wouldn't that mean I should see it as a benefit when someone has opposing views from me? It's possible I'll find an area to grow in! Also, I remind myself that my salvation depends on how I respond to being wrong/sinning. My salvation doesn't come from never being wrong or from no longer sinning, even after I'm born again. Being right all the time is God's burden to bear, not mine!

Contrastive thinking is the healthiest thing for the brain! Can you come up with **one** way a deeply held belief **could** be wrong? You don't have to agree or admit it is wrong, just come up with a way it could be. Having someone else give you the way won't help you. The act of thinking in this way essentially back-flushes the trees (dendrites) of the brain and keeps you mentally healthy. As a dad, one of the most important things for me to do in training my son to become a man is to help him be okay with, and quick to **admit**, when he's wrong. I mainly do this by quickly admitting when I'm wrong in front of him.

People who have a hard time with being contrastive on a deeply held belief ought to write down a belief, pause, and then pretend their enemy stated the written belief to them. Remember, we are naturally contrastive on everyone except ourselves. It is easy to look for weaknesses when we think others have stated the belief. Not only will this help your brain become healthier, it'll also help you be a better listener and sympathizer to others when they share opposing views with you.

Another "Weed Killer" Example from Scripture

Acts chapter 5 includes the story of Peter and the apostles being put in prison for their preaching and after their miraculous escape, they continued to preach. When charged by the high priest to not teach in

Jesus' name, Peter's and the other apostles' reply was, "We ought to obey God rather than men" (Acts 5:29b).

Where the "weed killer" moment comes in is when Gamaliel, a Pharisee himself and a doctor of the law, proclaimed to the religious leaders, "And now I say unto you, Refrain from these men, and let them alone: for if this counsel or this work be of men, it will come to nought: But if it be of God, ye cannot overthrow it; lest haply ye be found even to fight against God" (Acts 5:38-39).

Notice, he's asking the other religious leaders to give up control over what these men are preaching. Gamaliel knows that the truth will prevail regardless of the vessel, because all truth belongs to God. If I fight the words of my enemy but I'm wrong and my enemy is preaching truth, I'm actually fighting God. Gamaliel proved he was more interested in truth than being "right" himself. This is not surprising considering that the Apostle Paul himself stated that Gamaliel was his teacher (Acts 22).

CHAPTER 4

Life

WE'VE SEEN THAT truth is vital to profitability. We've also seen profitability is God's universal measure for progress on our journey. In the previous chapter, we looked at ways to handle distractions we all face that hinder our progress.

Jesus said, "I am the way, the truth, and the life" (John 14:6). We can see why Jesus is the key to progress: He is the truth. Furthermore, the definition in this verse for the word "way" is "progress, road, means, or journey." Jesus is the journey! Our ability to make it to the party depends on continual improvement in our interactions with Him. We need to let Him continually direct our actions. It is a process, not a one-time event.

We will conclude this part with an understanding of life. What is the definition of life? There are actually two popular usages of the word "life." They are both correct usages, but they cannot be used interchangeably.

The first use of the word life is "existence." When people talk about, "You only go around once in this life," they are talking about "in this existence on earth." This is a correct usage of life, but it doesn't apply in all cases.

For instance, when Jesus says we will have eternal life (John 3:16), does He mean we will exist eternally? Yes, we will, but people in hell will also exist eternally. The eternal "life" for those in heaven must mean something different than just existing. When I ask people to define life, I get a quick understanding of whether their view is one, two, or three-dimensional. What is your definition of life? What is your definition of death?

Three-Dimensional Definition

Some people have a one-dimensional view of life. They see it as a binary concept. Either you are physically dead or physically alive; however, Jesus said, "I am come that they might have life, and that they might have it more abundantly" (John 10:10). Again, Jesus can't just be talking about abundant existence. Clearly, there is another definition for this word.

C.S. Lewis wrote in *Mere Christianity* that the difference between a live body and a dead body is that a live body can, when it is damaged, repair itself to some extent.[8] Regarding non-contradictory definitions, you've probably noticed that I define words simply. I've found the most powerful and least contradictory definitions are the simplest. Let's look at life as "the ability to repair."

This would agree with Jesus because you could have, more or less, the ability to repair. What does it take to repair? The body must be able to fix damaged areas. In order to be able to fix these parts, the body needs the ability to create. Again, we are consistent with our model of good and evil. Evil destroys and good creates. Evil kills and good repairs. How does good create (repair)? Good does it with truth. So life, the ability to repair, depends on truth.

Jesus said He is the way, the truth, and the life. Now we know why these are tied together! If He is the ultimate access to truth, He is the ultimate creator (repairer). The ultimate creator would be the ultimate source of life.

What we've seen is a two-dimensional view of life. According to Jesus, you can have more or less "life." Jesus was not just speaking of physical life—the ability to repair physically. Jesus also said, "The words that I speak unto you, they are spirit, and they are life" (John 6:63). How can words **be** life, not just bring life?

Words can repair damage. Isn't this consistent with our definition of grace? There must be life in areas other than the physical.

The verse says life also applies to the spirit. In fact, we can take this a step further. The Old Testament says, "For the life of the flesh is in the blood" (Leviticus 17:11). If eternal life is the ultimate ability to repair, then it can only come through the ultimate blood. That ultimate blood belonged to Jesus! Now that we see our model is getting tighter, let's finish looking at the dimensions of life.

Life applies to emotional, spiritual, and mental states as well as physical. This is the three-dimensional view of life. Jesus offers us the ability to repair ourselves emotionally, spiritually, mentally, and physically, not only in this existence but also for eternity. This ability is necessary for us to continue the journey after we physically die.

It appears we have arrived at a more descriptive title for our two philosophies. Looking at the Profitability philosophy, it now looks like we can more accurately call this the "Life" philosophy. The Life philosophy begins by intentionally consuming resources in order to produce more. Likewise, we saw that "life" occurs once something is damaged (consumed) and needs to be repaired (created). The Life philosophy ultimately results in abundance; resources are seen as infinite. The Survival philosophy says nothing about abundance.

Survival Philosophy
1. Resources are finite
2. Humans consume resources
3. Minimize consumption to slow the inevitable: death
4. Long Term: Become more efficient

Life Philosophy
1. Resources are infinite if humans produce
2. Humans consume resources
3. Short Term: Use resources to generate resources of more value
4. Long Term: Become more efficient

Eternal Life

Finally, let's revisit the issue about eternal life and look at the implications. We have seen that people who get to continue the journey will have eternal life. This means they will have the ability to repair themselves for eternity. Therefore, the definition of death is "the inability to repair." This ability to repair is dependent on the ability to create, so people continuing the journey will be able to create.

Creation depends on truth, so people continuing the journey will have access to truth. Truth has been defined as facts that create, so people will be able to put facts together while they are continuing the journey.

What about people who aren't continuing the journey? It appears they won't have the ability to repair themselves, even though they will exist for eternity. They will be in Survival mode. In order for them not to be able to repair, they will not have the ability to create. This implies they will not be able to use truth.

But these people will know truth. The Bible says every knee shall bow and every tongue shall confess Jesus as Lord (Philippians 2:10, 11), yet they won't be able to continue the journey. They will have knowledge of the truth. How will they be prevented from using truth?

What does it take to use truth? It takes thought. A person must choose to think. How will people be prevented from thinking about truth and creating? Remember, people have a free will. They can choose to think or not to think. God can't impose His will on these people. God can't remove their brain or make their soul cease to exist because the soul is eternal. God can't wipe out their memory. This wouldn't be just. How can God justly prevent these people from thinking?

They will be tortured to the point they won't be able to think. Is this just? God uses justice to give these people what they have chosen. They didn't choose to value thinking during their time on earth, so they can be put in an environment that doesn't allow them to think.

We have already discussed the truth that God is just and can't allow a person to continue the journey if they do not want to keep going. This would violate their will. People must show their will to continue the journey through their actions. They must express it and choose to let God work through their actions. They must do the process.

The basic skill needed in this process is thinking. We are beings of volitional consciousness. We can choose to think or not to think. When people choose to understand, they are expressing with their will that they want to pursue truth and grow rather than be comfortable.

How do people stay comfortable? They refuse to think. This is done in many ways. If a person's actions aren't driven by thought, then what causes their actions? All that's left is their instinct and emotions. The Bible calls this the flesh. This refusal to think (understand) brings judgment on them and the only just way for them to spend eternity is not thinking.

Life Model	Hell Model
1. The ability to repair	1. Won't have ability to repair
2. Requires the ability to create	2. Won't be able to create
3. Choose to think and use truth	3. Won't be able to use truth
4. Physical, mental, emotional, and spiritual	4. Can't think because of the torture
5. Applies to this existence and eternity	5. Didn't choose to think on earth

Summary

When Jesus said He is "the way," He was saying that He is the means or the journey. We've been calling our goal of improving our relationship with God "the journey." We have seen that it is a process. This process involves truth and the ability to repair. We are the branch that gets pruned when we produce so that God can continue to flow through us and produce more.

God's universal measure is profitability. Profitability is ours when we express our will by choosing to pursue growth over comfort. Profitability is ours when we let God divinely influence our heart with truth and we choose to let this value reflect in our existence on earth. This intentional process requires thinking more, not less. Now we know why our success in getting to the party depends on understanding Jesus more, not less.

For the rest of the journey, we will measure everything using profitability to ensure our progress toward the party—that is, to make sure we are "good."

We now have the background necessary to add to the Reward Model and understand day-to-day Christian living.

- What are the ways to repair mentally?
- What are the ways to repair emotionally?
- What are the ways to repair physically?
- What are the ways to repair spiritually?
- How will this ability to repair benefit us after we physically die?

Joel Swokowski's Commentary

In the New Testament, we are often limited in our understanding of the meaning of many words due to the English Language. For instance, the word "life" is translated from multiple Greek words. The two main words that translate into life are "bios" and "zoe."

"Bios" is where we get biology. This refers primarily to our existence on earth as physical beings. It is often used in reference to the limitation of our "lifespan" , a specific amount of time, or "lifetime."

"Zoe" can also mean physical life but extends to the spiritual life we all have. Humans are physical and spiritual beings. Humans have a soul. That would be the zoe life within us all.

We can see that bios life is really the One Dimensional life presented in this chapter while the zoe life is the Second and Third Dimensions.

Now that we've been introduced to the definition of life, allow me to bring even more clarity to what "repair" means by contrasting it with "restoration." Repair makes things better than the original; better than before any damage was done. Restoration brings things back to their original state. Think of a classic car. Restoring that car would be fixing it to the point where it's the same as when it came off the line at the automobile factory. Repairing a car would be like taking your car that had engine problems and putting a newer engine in the car than it originally had. Now the car is even better than before it broke down, even better than it was originally.

So, the conjunctive definition for life is the ability to improve (fix) something to a better condition than it originally had. The freedom is improving (fixing) something and the limitation is that it has to be better than its original condition.

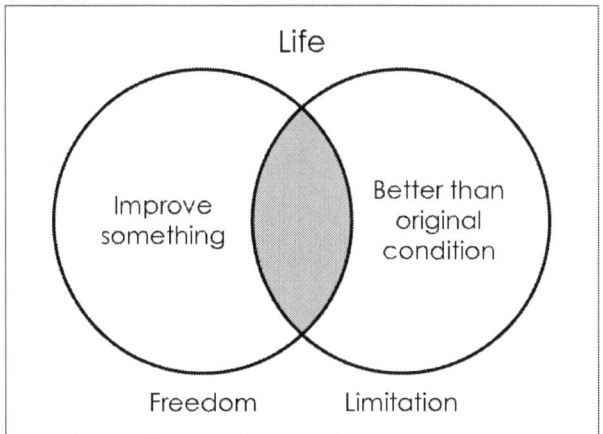

CONJUNCTIVE ILLUSTRATION FOR LIFE

When we're dealing with repairing the issues we have with people, even the spiritual, emotional, and mental issues, we can take the "bad" things that happen and repair them. Repair with people comes through confession and repentance. I see repentance as repair that is done as an effect of having done something wrong. This helps us understand the measure for repair: when the people involved are happy that the "bad" thing happened because now that they have repaired, their relationship is better than it was before the "bad" thing. This means that when we handle the issues we have with each other according to God's instruction (confess & repent), the situation can be made good.

Again, repentance is repair as an effect of having done something wrong. We can also repair proactively. We do this by continuing to grow in grace and faith. I don't have to do something wrong to repair. Repair will come as a result of allowing God's Spirit to live in and through my life. This is a picture of what will happen for eternity. Eternal Life would produce an increasingly better person. Eternal Death would produce an increasingly worse person.

In the next chapter, you will see yet another way in which you can intentionally and proactively grow the profitability of your life.

PART TWO

Pray & Love
(Exchanging Value)

INSIDE

Chapter 5: Prayer	254
Chapter 6: Prayer Implications	268
Chapter 7: Morals vs. Ethics	281
Chapter 8: Are, Do, Have	291
Chapter 9: Moral Code	301
Chapter 10: Determining Your "ARE"	306
Chapter 11: Love	322
Chapter 12: Love Applications	333
Chapter 13: Marriage	346
Chapter 14: Marriage Applications	355
Chapter 15: Living the Life Philosophy	366

CHAPTER 5

Prayer

AT THE BEGINNING of the first book, we asked the question, "How do we get closer to God?" Recall the directions we received were to believe, pray, love, and read the Bible. So far, we agree. We saw that we need to believe; that is, we need to have faith. Then once we understood what faith is, we found out why we need to believe. We are now able to do the "right" thing for the "right" reason. More importantly, this caused us to know how we can intentionally build our faith.

We found faith is the belief in causes we can't see or results that haven't happened yet because of understanding and experience. We know we can build our faith if we increase our understanding of and/or experiences with God.

We found that grace and uniqueness are universal principles. We also found the universal measure for intentional progress is profitability and is dependent on truth and the ability to create. Now that we have our universal measure of progress, we will see why the directions we are given are accurate. The background we got in the last section showed us how value is created. The rest of this book will examine how value is exchanged during every interaction that occurs.

This existence is simply a place for us to become more profitable by exchanging value. We can exchange value with God and others. The second portion of this book looks at how we exchange value with others. First, we will look at how we exchange value with God.

One way of exchanging value with God is through confession and repentance; however, this transaction is reactive. It is in response to mistakes we make. It isn't intentional in that we don't intentionally sin in order to exchange value with God. There is another way to exchange value with God and it can be done intentionally. It is called prayer.

Prayer

Simply put, prayer is conversing with God. This should involve not only speaking but listening as well. What is the reason for conversing with God? Basically, it is done in order for the individual to obtain a value. Whether it is peace of mind, an answer, or a miraculous healing, it all can be summarized as a value to the individual.

That is why most people say prayer is asking God for something. Now, we know we can't violate justice in order to get a value from God. So how can we get a value justly from God?

Remember, we have free will; God does not. We can do anything; God cannot. God can only do that which is righteous and just. Furthermore, God cannot violate our will. Even Jesus asked what their will was when He healed people. In fact, His ability to heal people depended on the person's will and their faith.

For instance, Jesus asked a blind man what he wanted (Mark 10:51). We might say, "It's obvious the blind man wants to see, why even ask?"

If Jesus had tried to heal the blind man and he didn't want to be healed, Jesus would have been trying to violate his will. We know he wouldn't have been healed, because all the successful healings in the Bible occurred after the ailing person expressed his or her will (in thought, word, and/or action) to be healed. Even though God knows our thoughts, He can't move until we express our will.

Prayer is the expression of our will and takes faith. In order for a prayer to be answered, we must express our will when we converse with God. This means that what we express must be something we have control over or it is witchcraft. If we ask God to do things that violate the will of others, then we are practicing witchcraft. People seem to have a hard time distinguishing between witchcraft and fantasy, especially in stories. Let's take a moment and look at witchcraft more closely.

Witchcraft

In the previous part, we discussed why fictional fantasy stories were better than "reality" stories at teaching truth. While a lot of fictional fantasy stories contain supernatural occurrences, they don't necessarily contain witchcraft. How can you tell if a story contains witchcraft?

If the story has a character that overrides the will of another in his life, then it is witchcraft. For example, turning another character into an animal or freezing him is something that is done to a person against his will. If the character in question expresses his will over inanimate objects to hinder another person, it is not witchcraft because we are talking about inanimate objects. The will of the other person is free to respond to the inanimate objects. Whether the story glorifies witchcraft or not leads to a different discussion.

Getting back to prayer, is it witchcraft to pray for others? If others ask us to pray for them, then they are expressing their will that we pray over their actions. This is not witchcraft. If we pray for inanimate objects or circumstances, then that also is not witchcraft. Think of Elijah or Jesus praying over the weather. But it is witchcraft to pray that someone believes something or does something when he or she is against it. This includes praying for people to get "saved" or healed. How are we supposed to pray?

Prayer Structure

Prayers in the Bible tend to follow a general structure. There are four phases, with the first one identifying that it is God we are praying to and praising Him. We have to state our will that our prayer is towards God and God's power through praise. Our focus is on God as the source. Remember, it takes faith to pray, and this is the beginning of our building that faith.

The second step is to reference something that proves our faith. Remember, faith comes through understanding and/or experience. This is a continuation of the faith-building process. We may reference an experience we had that proves God's ability or we can reference understanding, usually a Biblical passage that proves God's ability.

We know the goal is to ask God for a value, which is the fourth and final phase. How can we justly ask God for a value? We need to bring justice into the prayer. The third phase of the prayer focuses on us either showing why someone deserves a reward or why someone deserves a penalty. Referencing justice completes the process of building faith. It serves as the final proof to us that we can expect to get a response from God.

Now, we are ready for the part we are all familiar with: asking for a value. The final phase of the prayer is to express our will and ask God for a value. This is where we have to be careful.

Requesting a Result

In the cases where we are requesting a punishment as the value, we can't pray for a specific punishment because that would be witchcraft. We pray for a result. First of all, it is best to pray for a result because God might have a better way than ours to bring about the result.

More importantly, the process of prayer finishes with us turning the situation over to God. So it shouldn't matter to us how God brings about the result. The decision to turn it over to God must be final and made by faith. When we turn the situation over, we are telling God we won't try to take justice into our own hands.

God doesn't move to invoke justice until we assure Him we won't do it. This shows God's mercy. If God did invoke justice before we turned it over, it would require us to get a punishment if we ended up taking justice into our own hands at a later date.

Prayer is the expression of our will with faith and has to be about things within our control. This would include things that may happen to us in the future that are seemingly random. For example, we can pray that we won't get sick; however, our prayer needs to be over things we can control. We can't control our exposure to disease, but we can express a will over our health. The prayer for this example should be to request a result that focuses on our health or protection.

This brings up another issue; we need to follow God's example and ask for what we want (e.g., health), instead of what we don't want (not sick). The Bible says God "calleth those things which be not as though they were" (Romans 4:17).

We are not supposed to ignore what exists. We can't call things that are as though they are not. We can't say, "There is no brick wall" and then try to run through it. When it comes to illness, the individual who is ill shouldn't be saying, "I'm not ill." He should say, "I am healed."

Prayer Model
1. Recognize God and praise Him
2. Build faith through experience and/or understanding
3. Reference justice
4. Express your will over something in your control
5. Ask for a result
6. Forgive and let it go

Summary

When we pray, we can choose to have our reward now or for eternity. We can choose to have a reward for ourselves or a value taken away from the offending party. We cannot pray over someone's will. If we are praying for someone, that person has to express his or her will to us in order for us to pray for him or her, and it has to be over something that is within that person's will.

Let's look at an example and some implications in the next chapter.

- How often do you just talk with God without praying?
- What is something you specifically want?
- Have you specifically asked God?
- Is it something within your control and not against someone else's will?
- What would hinder God from giving you what you are asking for?

Joel Swokowski's Commentary

The rest of this book is focused on exchanging value. We need to keep this in mind!

Prayer is second only to God's Will as the most complicated doctrine in Christianity because it relies on all of the doctrine we've covered so far. In order to have a non-contradictory explanation for prayer, you would need to have a non-contradictory explanation for every doctrine we've covered so far. If even one of these doctrines is contradictory, you won't be able to explain prayer in a manner that explains every instance of prayer in the Bible.

This is just one reason this chapter brought much persecution against the author. Pastors and leaders once again proved they were either not reading the book or they were misinterpreting it. Both of those issues weren't wrong in and of themselves until those same pastors and leaders then judge the author as a heretic without first getting more understanding. Allow me to bring some clarity to this very complex doctrine of prayer.

You can talk to God without asking for something (and we all should do this). It's just not prayer. Many people consider any conversation they have with God as if it's prayer. If that's your definition, I understand your confusion as to defining prayer as "an exchange of spiritual value." You may be thinking, "I'm not wanting anything from God, I'm just wanting to share and fellowship with Him." Great, and again I would say, you should do this. What I've done to help people with this conflict is bring them the resolution of fellowship or prayer. For now, let's look at the growth of our intimacy with God as "fellowship" while any communication that involves "asking" for something from God then means that communication has become a "prayer." It makes more sense when you understand that the word "pray" simply means, "to ask."

This chapter began covering how to exchange value with God through prayer, and this presentation was an overview of prayer. To bring more clarity to this overview of prayer, let's look at the approach two different teenagers use to ask their parents for $50, and then I have a question:

> The first teenager casually asks his parent for $50 and in the middle of his request, he suddenly stops to respond to a text message from his friends. When the parent asks why he needs the money, he begins to demand the money and even insults the parent. When the parent reminds him he was given $20 last week and ended up using the money to buy alcohol and get his friends in trouble, he plays the victim saying, "I knew you wouldn't give it to me because you don't love me." When the parent tries to explain that what they are doing is actually love, the teenager leaves the room.

> The second teenager picks a specific time to approach the parent in an intentional manner. The teenager makes sure there are no distractions and carefully makes his request in

a humble and respectful manner. The teenager explains why he needs the money and how he deserves it (e.g., based on chores he has done) and how he spent the $20 from last week to benefit others. In fact, the teenager is able to show that the $50 would accomplish things in line with what the parent has wanted to occur. Furthermore, when the parent explains the same effect could be had with just $30 **or** with $50 not only could the teenager accomplish what he needs but he could also do something more for the parent, the teenager listens and agrees with the parent's plan.

Here's the question: Why would a person treat the God of the universe in any manner less than the way the second teenager treated their parent?

Is Prayer Free?

I have two wonderful nieces who brighten my day every time I see them. I'm the "candy uncle" due to the consistency with which I give them candy. I've definitely enabled this nickname! I love the sound of their voices when they run to me and yell, "Candy!?"

Recently, Layla (the oldest and 6 years old at the time), asked for two pieces of candy. I had to reject her because I had run out! She was confused. It was as if she was thinking, "But you're the candy guy!" In fact, she did directly ask me, "What do you mean?" I shared with her that I was out of candy and told her I needed to buy more. She was still confused, as if because I'm the "candy uncle," I would always have candy. A contradiction was introduced. How can the candy uncle not have candy?

Her parents are amazing, and like the author of Modeling God has done, they've been teaching their girls the principles within this book from as early on in their lives as possible. I took an opportunity to teach Layla a "justice" lesson. I asked her, "Where do you think the candy comes from?" Her response was, "It's FREE!"

Wow! This is amazing! Now I saw the reason and the value behind my question. In her view of reality, candy **was** free. I even said, "Yeah, it's free…to YOU!" I then shared with her that just because she doesn't see the cost being exchanged between her and me doesn't mean there isn't a cost. The candy is not free. Not to me. Not to her. She may not pay for the candy the moment I give it to her but justice says that every time I give her candy, it does cost something.

Now, the value I receive from the joy I experience interacting with Layla and Rosie makes me feel like I owe them. I'll continue to give the candy and I'll continue to not charge them for it. I feel like I'm still coming out ahead! However, this was a great time for me to help Layla understand justice and value exchange.

This entire exchange with Layla caused me to question my own behavior, especially as it related to my interactions with God. How often am I interacting with God in the way my 6-year-old niece interacts with me? Do I recognize the gifts God has given me? Do I recognize the cost of those gifts, even if it wasn't me who paid the price?

This reminds me of the infestation of the "one-way justice" culture we live in. I don't count the cost of the things I want to receive. I only count the cost of the things I have to pay for. When people say, "God's Nature is Love," and, "Grace is unmerited favor" they assume, "this must mean God wants to just give to me and it's free!"

Think about it, even though God's gift of Salvation is free, that doesn't mean it didn't **cost** anything. It's a free gift in that I do not and cannot do any work to earn or pay for it. Jesus did the work! Jesus paid for it!

What would Jesus say, or how would He feel if I told Him that I didn't think my salvation cost anything? It would be like saying, "But it's FREE, Jesus!"

I wonder if Jesus would feel like I'm ignoring the massive sacrifice He gave to pay for my life. I wonder if Jesus would think, "Well, it didn't feel free to Me!"

God is a good Father who gives us good gifts, and because He is a good Father and because He is Right and Just, we know that prayer is **not** free. It is an exchange of spiritual value between us and the Creator! What an amazing privilege we have to exchange with Him in this way. Shame on me for taking that for granted and for minimizing the cost of the gifts that I've been given!

It seems that most people think that prayer, or more specifically the answering of prayer by God, is free. Where is this belief rooted? I think the answer to this goes to whether or not the person has an accurate understanding of God's Nature. If God's Nature was love, then yes, He could just give and give and give…and it wouldn't cost anything. But if that's the case, it introduces loads of contradictions: why did He have to send Jesus to die if He could just **give** us something for free? Why wasn't Jesus wasn't able to heal certain people? For justice reasons, including their lack of faith. This further proves how witchcraft doesn't work, how praying over someone that isn't in agreement with the prayer doesn't work. If Jesus could do witchcraft, all of His prayers and healings would have worked perfectly.

God's Responsibility vs. Believer's Responsibility

There are certain aspects of our lives that God and Jesus have direct control over and others they don't. Matthew 15:38 says Jesus couldn't do many miracles in His hometown because of the unbelief (lack of faith) of others. In Matthew 9 you can see that Jesus healed everyone of every illness and they were still struggling mentally and emotionally.

After Jesus healed everyone of every manner of illness, He had compassion on the people who were distressed and scattered, seeing them as sheep without a shepherd, and it grieved His heart! The meaning of "distressed" and "scattered" speaks to the people being cast down in their thought processes and emotionally disconnected.

Jesus explained and encouraged the need for "laborers" because in order for people to be restored mentally and emotionally, it requires interacting with people! Through the story in Matthew 9, we see that God and Jesus are responsible for restoring a person spiritually and physically, which can be done instantly, while the church is responsible for restoring (and even repairing) a person mentally and emotionally, which requires hard work (hence the term "labor").

God can instantly respond to our spiritual and physical needs. Everything that is mental and emotional needs to be dealt with by us, however, others (including God) can speak words that help us mentally and emotionally.

Who are you praying to?

Again, I will stress the importance of the resolution between the *what*, *why*, and *how*. What's more important: knowing who the person you're talking to is in their causes (*how/why*), or stating the right name (*what*) to the person you're talking to?

Identifying God as the person we are praying to isn't just saying the name (*what*) God or Jesus. You can say God or Jesus and pray to something else. You can not say God or Jesus or use another name and state something that is only God: the Creator, etc. This is why knowing God is right and just (*how*) helps a person's prayer life.

The Four R's of Prayer

This may seem like a lot to remember. You may be thinking, "Joel, I just want to be able to talk to God without remembering all this stuff!" I get it. I also understand that the better I know how to intentionally talk to the God of the Universe, the better my fellowship with Him, and the more likely it is that my prayers will be answered. Is the God of the universe worthy enough of your time for you to learn the best and most respectful way to interact with Him?

If you remember nothing else about the author during this chapter, remember this: the author held prayer in such high esteem, he approached it with the utmost respect. Is it possible he was attacked because the religious leaders realized they not only didn't understand prayer, they weren't giving the God of the Universe the utmost respect? Finally, remember that Jesus said there was a wrong way to pray in Matthew 6, so there is a right way to pray.

Go ahead and look back at the Prayer Model. I want to give you a way that might make it easier for you to remember what your responsibility is as it relates to requesting a value from our Lord. I call them the Four R's of Prayer:

1. **Recognize** God.
2. **Reinforce** your Faith.
3. **Reference** Justice.
4. **Request** and let it go.

CHAPTER 6

Prayer Implications

THERE IS ANOTHER dimension to the Reward Model (Figure 8). When Person A takes a value from Person B (Panel 1), justice equals everything out by giving Person B a spiritual value and removing a spiritual value from Person A (Panel 2), so rewards are occurring on two levels simultaneously. Prayer initiates the exchange of this spiritual value into our physical existence. Person B can pray a value down to himself (Panel 3) or he can pray the loss of a value down to Person A (Panel 4). Either of these prayers may require additional value.

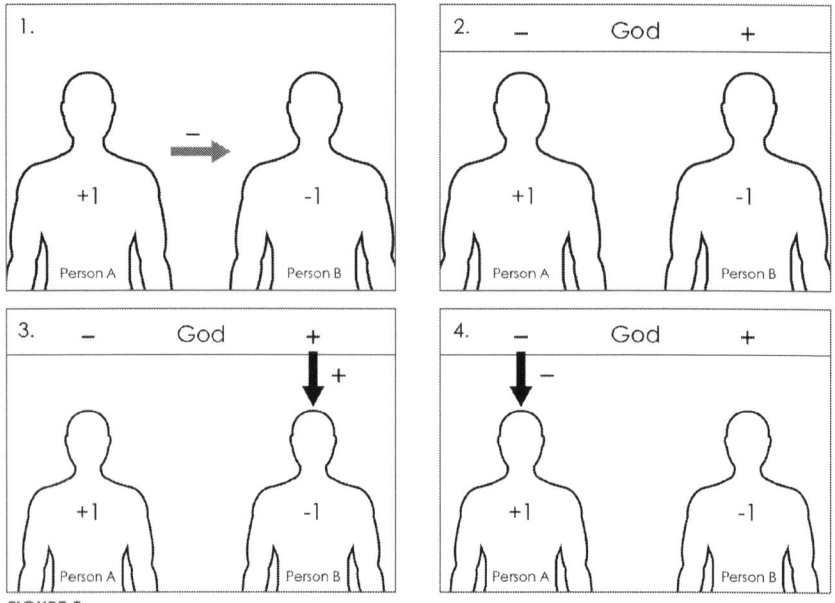

FIGURE 8

Example

I have selected a relatively short prayer below. As you read other prayers, notice they differ in the length and depth of the first two phases, but they all reference justice and ask God for a value.

> "And Hezekiah prayed before the Lord, and said, O Lord God of Israel, which dwellest between the cherubims, thou art the God, even thou alone, of all the kingdoms of the earth: thou hast made heaven and earth.
>
> Lord, bow down thine ear, and hear: open, Lord, thine eyes, and see: and hear the words of Sennacherib, which hath sent him to reproach the living God.
>
> Of a truth, Lord, the kings of Assyria have destroyed the nations and their lands,
>
> And have cast their gods into the fire: for they were no gods, but the work of men's hands, wood and stone: therefore they have destroyed them.
>
> Now therefore, O Lord our God, I beseech thee, save thou us out of his hand, that all the kingdoms of the earth may know that thou art the Lord God, even thou only" (2 Kings 19:15–19).

The first sentence actually contains the first two phases of the prayer. The first sentence consists of Hezekiah recognizing God and praising Him as the God of all the kingdoms. Hezekiah builds his faith through understanding by recognizing that God made heaven and earth.

The second sentence makes a request of God to hear the words of Sennacherib. The third sentence invokes justice. Hezekiah says that the kings of Assyria have done wrong by being destructive and worshiping other gods.

The fourth sentence requests a value. He wants to be delivered out of the hands of the people who worship other gods. In this case, Hezekiah wants a value given to him and his people. Notice, he didn't tell God how to save them. He prayed for a result. Also, Hezekiah could have prayed for a penalty to fall on Sennacherib. Instead, he used justice to request a value.

Implications

The most important implication of prayer is that we have to be aware of where we stand in relation to justice. Notice, this is how most prayers go wrong. If you ask for a value without having enough justice for yourself (Figure 9), then how can God justly grant a value?

If you have spiritual value (Panel 1), your prayer is an expression of your will to bring some (Panel 2) or all of that spiritual value down. Here, the only way for you to get value is to have handled justice well. Notice, this process doesn't create. It is only an exchange of value.

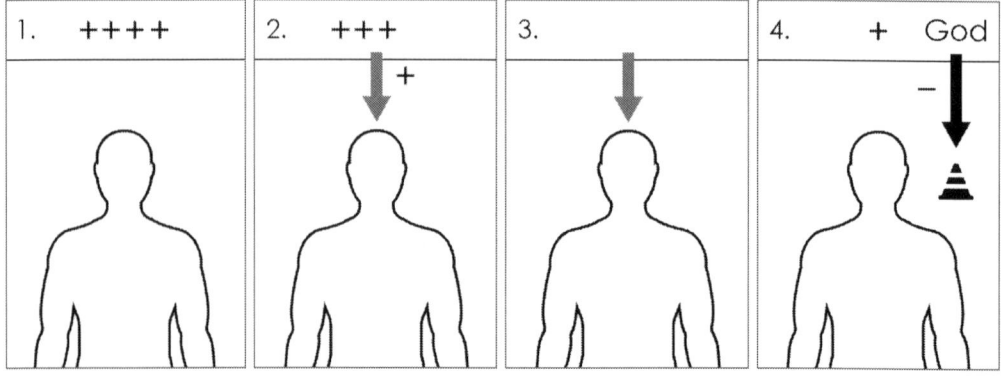

If you don't have value (Panel 3), God would have to first take a value away from you before giving it back in answer to prayer (Panel 4). God does this by putting an "obstacle" in your way. If you handle the obstacle well, then you will get a value. If you have a bad attitude or don't handle the obstacle well, then you have forfeited your reward and may even end up with a greater deficit (Figure 7, Panel 4).

You've probably heard people say they no longer pray because every time they do, bad things happen. Well, if you are praying for a value and you don't reference justice or have enough spiritual value, having "bad" things happen is the only way for God to justly give you a value.

In fact, when a church tells its members to pray for a collective cause, the church is relying on the members to have justice on their side. They are counting on their members to have handled adversity well. If the congregation doesn't have enough value, then the church and its members are going to have to handle adversity after they pray in order to get the value.

Prayer initiates an exchange of value with God because it is an expression of your will. If you have no value due to you through justice, then God creates a situation for you to earn a value. You have expressed your will by asking for a value. This act essentially gives God your consent to give you an "obstacle."

There are two other ways for you to get value from God. First, you can forgive by turning over situations you have been holding on to from the past. If someone has wronged you in the past and you haven't obtained your justice, you can turn the situation over to God and receive the value.

The other way is to do things that intentionally build value. Giving people a value they can't or won't pay back does this. We saw this earlier when we understood why we should be kind to our enemies. When you are kind to your enemies, you are acquiring a value through justice because your enemies are a sure bet not to respond by giving you a value. In fact, being nice to your enemies is the best way to remove your enemies. When you give them a value, it will cause them to be meaner to you, and that allows God to punish them (Figure 7, Panel 4).

The other action involves giving a value to people who can't pay you back. This involves the poor, people in prison, etc. Whether it is your enemies or people who can't pay you back, when you give the value, that is the last action you take. You can't brag about it or hold it over the person's head. You can't bad mouth the person or complain. Otherwise, you are getting the value back before it can come from God. Remember, these actions have no effect on salvation!

You are told to pray. Yet, prayer doesn't really work unless you have value. Perhaps the reason you are supposed to pray is that it requires you to do things that give you spiritual value. It also requires you to handle justice well. When you understand the implications of exchanging value, several passages of the Bible take on a deeper meaning. For instance, this is just one reason why Paul says he glories in tribulation because it is an opportunity for him to obtain spiritual value (Romans 5:3).

Another passage says that if you have a problem with your brother, you should leave the altar of prayer and fix it before continuing your prayer (Matthew 5:23, 24). Other passages say having unconfessed sin (Mark 11:25), harboring anger against another; especially a spouse (1 Peter 3:7), etc. actually hinders the prayer. This is because justice is against the individual. They aren't handling justice well. Praying for justice when you have a penalty due in another area is only going to result in a value being taken away.

You might have noticed that the meanest people seem to get by in this existence without a penalty. We saw earlier (Figure 7, Panel 4) that part of the reason is that they are so mean, the people they wrong end up taking out their own justice. Whether it is physically, verbally, or through their emotions, the wronged individuals are blocking God, during this existence, from acting on their behalf. If the offending person was less mean, we would probably let the action go and God would invoke justice.

In addition, the offending person needs to forgive in order to get their value. Isn't this exactly the type of people we see get rewarded in this existence? Throughout history, people have wondered why God allows people to prosper who are obnoxious and oblivious. Being obnoxious causes them to offend others. Being oblivious causes them to forget the people they've offended and not be able to retaliate when the offended take a value from them.

Again, the quickest way to make sure your enemies get their justice is to forgive them and turn it over to God. Don't strike back. Overcome evil with good. Likewise, our prayers should conclude with us turning the situation over and letting God determine the specific result.

Praying About Others

I mentioned that praying for someone to get saved could be witchcraft if it is done against his or her will. How should we pray for family and friends to begin and continue the process of salvation? Remember, we can only pray over things for which we have control. Also, remember that we need to have enough value to account for our request.

Actually, there are several ways. For example, you have control over yourself. You have control over the words you say and your actions.

God knows the people and what they uniquely need to hear and see. One way would be for you to ask God for the words these people need to hear and/or the actions they need to see.

Another example would cover the situation where these people aren't talking to you or able to see you. You could pray that God would give the words to a person who is currently telling God they would like to be used by Him to spread His message.

These are just two examples. There are obviously many more. Like all models, what I've shared are principles. How they apply to each unique situation still requires thought on your part and has various levels of success depending on you.

Let's look at a quick example from the Bible of someone praying incorrectly over someone else. David prayed an interesting prayer in 2 Samuel 15:31, saying, "And one told David, saying, Ahithophel is among the conspirators with Absalom. And David said, O Lord, I pray thee, turn the counsel of Ahithophel into foolishness."

David was battling his son Absalom at this point. David's lead advisor was Ahithophel, who was now advising Absalom. David prays over another person (Ahithophel). More specifically, David prays that another person's counsel is turned to foolishness. Let's examine the prayer relative to the four structural aspects.

First, David recognizes he is praying to God when he says, "O Lord." David got the first part right. The second step is to build faith by recognizing something God has done through personal understanding or experience. Nothing is recorded that says David did this faith-building step during this prayer. The third step is to reference justice. Again, nothing is recorded that says David referenced justice during this prayer.

In fact, we have to further question whether David was on the right side of justice. The fourth step is to make a request. It looks as if David did this. However, remember the request has to be about something within your control or will. Is Ahithophel's counsel within David's will?

Isn't this witchcraft?
Isn't David making a request over someone else's will?
What is the result of the prayer?

"And the counsel of Ahithophel, which he counselled in those days, was as if a man had enquired at the oracle of God: so was all the counsel of Ahithophel both with David and with Absalom" (2 Samuel 16:23).

David's prayer was not answered. In fact, the opposite result occurred. I believe this story shows what happens when someone who even has a heart after God participates in witchcraft. They don't get what they want.

Summary

When you pray to God, you are really saying you have given up the idea of getting justice with your own might and now rely on God, through faith, to bring justice. This requires that you let go of the hurt, scheming, and bitterness you have; otherwise, the prayer won't be answered.

There are three ways to get a value from God:

1. Handle adversity
2. Turn a past injustice over to God
3. Do something for someone who can't or won't pay you back

When you have been wronged, you have a value coming to you. You have two options. First, you can determine who you want to exchange value with God; you can receive a value or the people who wronged you can receive a punishment. Second, you can determine when you want that value exchanged. Remember that you don't have to use that value here. You can have your reward in this world or eternity, but you must express your will. God can't move until you intentionally state your objective. Finally, we have added to the Reward Model by realizing that prayer initiates an exchange of value.

Reward Model
1. If a person gives a value, justice equals it out in spiritual value
2. If a person takes a value, justice equals it out in spiritual value
3. God won't deliver you a value until you forgive or forget
4. Prayer initiates an exchange of value

Remember, making the reward scriptures a requirement of salvation is called "legalism." Different models govern salvation and rewards.

- Identify a prayer you believe was not answered by God.
- What are the possible reasons this prayer wasn't answered?
- What would you do differently?
- How have you handled an injustice well?
- How have you turned over a past injustice to God?
- What did you do for someone who can't or won't pay you back?

Joel Swokowski's *Commentary*

This chapter brought even more oppression against Lenhart. I've seen and heard statements like: "The author teaches that people ought to be selfish and do things for their own gain, against the words of Jesus who teaches selfless behavior," and "Lenhart teaches that you shouldn't pray for other people to be saved."

These are both gross miscalculations. Allow me some time to help us understand the prayer implications a bit more clearly.

During His premiere teaching (Matthew 5-7), Jesus spent a large portion of that sermon teaching us how to intentionally gain value/reward. Matthew 6 in particular was blatantly explaining ways for us to do things for others that would result in us gaining! Isn't this the definition of selfishness? Well, maybe not. Let's look at four terms to help us understand Jesus and what He was teaching:

1. Selfishness: This is the perspective that focuses solely on the individual at the expense of others. This type of person has no problem hurting, abusing, or bullying others to get ahead in life.
2. Selflessness: This is altruism. It is the belief that your brother's needs are greater and better than your own and it is your duty to sacrifice your own happiness, value, and ultimately self-esteem for the benefit of everyone else. True selflessness is impossible and would only result in one's death as there will always be someone hungrier or more in need of that water than yourself.
3. Self-Interest: This is what is called the American Dream. Everyone has the right, as long as it doesn't impose on others' freedom, to create an abundant life for themselves.

This perspective is focused completely on one's self. Self-Interest is focused on the immediate present and upfront appears to be right. However, due to this behavior's focus on the short term and the physical, it ends up being the most destructive approach. This is because of its deceptive nature: It appears right (right *what*), but is not a spiritual focus (wrong or no *how/why*).

4. Rational Self-Interest: This perspective is what is known as interdependence and is a conjunctive.

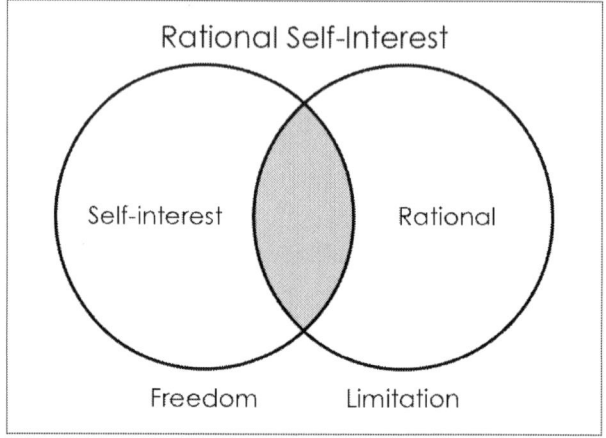

CONJUNCTIVE ILLUSTRATION FOR RATIONAL SELF-INTEREST

The rational half of the conjunctive is long-term focused. A person who has embraced this perspective knows they are not complete. They are not fully right and not fully just and they will never be fully right or fully just. Therefore, they are in need. This person would do things to intentionally grow themselves in the spiritual and long-term capacities. For instance, your salvation! Who directly benefits from your salvation? You do! You are also storing your treasure in Heaven, as Jesus encourages us to do. This is a long-term focus and a spiritual focus. We are going to see that Jesus **only** used this approach and the result of this approach is that others benefit as well (hence the term *interdependence*).

Praying for others

The more I've grown as a prayer warrior, the more often I see people flippantly requesting prayer from the masses. Think of the times you've seen someone post on some social media platform, something to the effect of: "Pray for me" or "Pray for my friend/family/etc." Is this wrong? No! Not at all. However, I wonder two things:

1. Does this person know that they are asking for value/reward from all the people who read this? Wouldn't this be comparable to my requesting money from everyone on my social media page? Again, not wrong, but as it relates to prayer, am I still seeing it as free?
2. What are these people requesting me to pray for? So often the request is simple, "Please pray." For what? I've found the best way for me to invest my spiritual value is for me to understand the specifics of the request, who I'm praying for, and what **they** want.

As it relates to praying for people's salvation, this chapter explains it very well. I do want to emphasize, Lenhart is not saying you shouldn't pray for another person's salvation. He's presenting information for you to help you make sure whatever you're praying for, that you're doing so in the way that God will actually be able to respond to your prayer. Do you want your prayers to be answered? Do you believe that the manner in which you make requests unto our Lord is at least as important as the prayer itself?

Finally, since this chapter spoke about exchanging spiritual value for physical things, I want to point out the "conversion" between physical and spiritual. In Numbers 12, Miriam speaks against Moses. This can be seen as a "spiritual" attack because nothing physically was felt by Moses. How did God equal out justice? Miriam got leprosy!

We tend to think a physical injustice is worse than a spoken/spiritual injustice. It is not. This supports what Lenhart said about bad people prospering: they do an injustice and gain much more from all the people speaking against them. While that ought to be enough proof, notice what the Bible lists as abominations to God. Proverbs 6:16-19 lists seven things and at most two are physical. The other five are spiritual including "he that soweth discord among brethren."

CHAPTER 7

Morals vs. Ethics

PRAYER IS THE only way to intentionally exchange value with God. In fact, it is the only way for us to get God to take an intentional role in our lives. Now we know why we are supposed to pray. Notice, we cannot become profitable during this exchange of value with God. Remember, the Salvation Model says we get a value from God and become profitable by exchanging value with others. God initiates this value.

The rest of this book focuses on how we exchange value with others. Every interaction between people involves an exchange of value. People are either gaining or losing with each interaction they have. The ultimate way for people to exchange value is through love. It is a complicated process and we will need more background before we define love.

It appears our actions have results, and those results can be profitable or unprofitable.

There are two kinds of actions that people recognize as profitable. Said another way, people recognize two possible causes for these profitable actions: morals and ethics. People seem to use these interchangeably.

For our purposes, we are going to define morals as "a set of principles that result in profitable actions in the big picture and long term." They are profitable actions regardless of the location or the era. Specifically, they are actions that would be profitable regardless of when a person lived or in what culture (where) they lived. For example, we can all agree that individuals should not abuse each other physically for their own personal benefit. This is pretty much true for every culture that has ever existed.

Let's look at the interactions between individuals more closely.

Getting a Value From Others

When individuals interact, they do so to exchange value and become profitable. We saw in the last chapter there were three ways to get a value from God. Now we are going to look at the three ways we can get a value from other people.

The three ways people obtain value from another individual are:

1. Threaten them physically
2. Appeal to their emotions
3. Give them fair value

Let's look at each interaction from the perspective of profitability.

The first way is to threaten someone physically. "Give me that or I'll hurt you." If I threaten you with physical harm, then I am requesting a value in exchange for not hurting you. Basically, I am asking for a value and promising you I won't take a value from you. The extreme case of this would be to put a gun to your head and threaten your existence.

When you give me a value, are you getting a value? Some would say yes because you didn't lose your physical existence. In this case, however, you had your physical existence before our encounter, so technically, you didn't gain anything. You didn't lose your existence and you lost the value I requested. This encounter results in my gaining a value and your loss of a value.

The second way is to appeal to you emotionally. "Please give it to me. Be a nice person, not a mean person." Some people call this begging. What I am really doing is asking for a value and promising you that you won't feel bad. Again, are you receiving a value? Some would say you are because you are not feeling bad; however, I would say you didn't feel bad before the encounter, so the non-loss of this value is not the same as gaining a value. If you give me the value I begged for, this encounter results in my gaining a value and your loss of a value.

The third way would be for me to offer something you valued in exchange for the value I was requesting. In the end, we both become profitable or we don't do the deal. This is the only moral way to interact with others. We must offer people something in exchange for what we want. Profitability and uniqueness allow us to do this in a fashion that both gain out of the encounter because we all value things differently.

Truth also played a role in the creation of profitability because we appealed to each other through reason. The first two methods did not involve truth and therefore didn't involve profitability. Actually, both of these non-profitable exchanges will get rectified in the end through justice. In fact, there is another aspect that distinguishes the differences in these ways.

Guilt

The first interaction accentuated the physical. The transaction occurred because I was threatening physical violence. The second encounter occurred because I was accentuating the emotional aspect. The third encounter occurs through logic—I had to think and appeal to you with reason. So we see there is a way for individuals to become profitable other than God initiating the transaction via grace.

This interaction is governed by justice (Reward Model) and has no effect on our salvation. How does justice allow two people to gain without having to pay for it? Profitability is a result of unique individuals exchanging value. Since God created uniqueness, He is still the ultimate source of creation. When this value is created, both sides are literally giving the other person half of the value. In that way, they are following justice because they are giving each other half the value and receiving half the value. There should be no guilt resulting from this exchange.

However, if there is an inequality in the division of the created value, then there will be guilt, and justice will eventually rectify it. For instance, recall the example in the Profitability chapter where you value an object at $15 and buy it for $10 from someone who values it at $5. If you had bought it for $8, you would have received $7 of the $10 that was created, while the seller would have received $3. Eventually, justice is going to require the extra $2 you received. Notice, this does not make you unprofitable, because you still achieved $5 of profitability.

Unfortunately, the first two ways always result in me feeling guilty because I know justice exists and I didn't give anything in return for the value I received. Subconsciously, everyone knows they can't keep getting value without giving a value. They know justice will eventually equal everything out. We all feel this inequality in justice through guilt. This guilt is a powerful force.

When people commit spiritual acts that cause guilt, this guilt is a warning. God built us this way so we know when we are acting apart from righteousness and justice. The conscience tries to let the individual know he needs help; that is, he needs to correct the action. When this warning is ignored, the conscience needs to increase the severity of the warning because the removal of the guilt is the ultimate goal.

Eventually, the warning manifests itself in pain. This is a warning sign to the individual that he needs help. The individual has the free will to act apart from his nature by pursuing growth and dealing with the pain or according to his nature by pursuing comfort and trying to ignore the pain.

If the pain is ignored, it begins to manifest itself in the next phase: fear. People frequently develop seemingly irrational fears. They may not even have a conscious idea why they have these fears. We already talked about how when a psychologist works with a person, they reverse the process. They identify the fear and try to find out when the person felt the pain. Then they try to tie the pain to guilt; for instance, he hated his mother.

The Bible clearly states fear is proof we have lost our way because God has not given us a spirit of fear (2 Timothy 1:7). So if we have fear, it comes from something apart from God. Taking what we learned in the last book, guilt hinders life; that is, our ability to repair and create. In fact, this ability to repair is what is needed to remove guilt.

Ethics

We saw that people try to remove this guilt through apologies. This is done on an individual basis. Ethics is another way people have dealt with guilt. It is different from an apology in two ways. First, its focus is not on the individual, but on society. Secondly, it doesn't even acknowledge the guilt. It tries to completely avoid this need to remove guilt. Society simply declares the immoral action that caused guilt to be allowable. Ethics are a set of principles that result in actions that are accepted as correct by a specific society for a specific location and time in history. Even though the actions result in one person losing profitability through coercion, the society agrees there is no need for justice to equal out this specific case.

Ethics is man's way of removing guilt for not being moral. Basically, groups of people agree it is okay not to be moral as long as we act according to an accepted, special code of conduct. The best example of this is business. People who aren't moral in their business dealings rationalize the guilt by saying they are following "business ethics"; however, the guilt still remains.

Notice, the difference in ethics and morals creates an opportunity for us to justly cause others to persecute us and give us value. This can be accomplished by doing something that is not immoral but unethical. We saw an example of this in the first book with Moses and Pharaoh.

Moses made a request of Pharaoh to allow the Israelites to go into the desert to worship God. This request was not immoral, but Pharaoh considered the request unethical. This caused Pharaoh to unilaterally act unjustly against the Israelites. God knew Pharaoh had an ethical view and would consider God's just (moral) retaliatory actions to be even more unethical. We saw how God's sovereign hardening of Pharaoh's heart was an effect of Pharaoh expressing his will.

This discrepancy between ethics and morals is also how Jesus justly acquired value. Jesus did things that were moral (e.g., healed on the Sabbath), but the leaders of that time considered these actions unethical. The leaders said the actions were against the law. Jesus was moral and unethical.

This is the same principle civil disobedience is built on. Basically, groups of people are supposed to do something that is not immoral but unethical. After they are persecuted, they need to take the abuse and forgive so that God can equal out the situation through justice.

Morals Model	Ethics Model
1. Never changes	1. Changes with location and time
2. Cause of actions	2. Cause of actions
3. Hinders us in short term	3. Allows us to accomplish things in the short term
4. Leads to long-term profitability	4. No thought is given to long term

- How do you typically obtain value from other individuals?
- How do you respond to people who threaten you?
- How do you respond to people who appeal to your emotions?
- How often do you give fair value?
- What are five things that are not immoral but are unethical in our society?

Joel Swokowski's *Commentary*

This chapter states that Jesus acted with morality, often not ethically. At least, we definitely see how the religious leaders of His day believed what He was doing was wrong (unethical). One of the ways we can know for sure that Jesus was moral was that He always gave a reason and a value when He asked for a value from someone. He always approached people with the third way of motivation: Giving them fair value. In the previous chapter, I called this "Rational Self-Interest".

For instance, take this passage: "Take heed that ye do not your alms before men, to be seen of them: otherwise ye have no reward of your Father which is in heaven. Therefore when thou doest thine alms, do not sound a trumpet before thee, as the hypocrites do in the synagogues and in the streets, that they may have glory of men. Verily I say unto you, They have their reward. But when thou doest alms, let not thy left hand know what thy right hand doeth: That thine alms may be in secret: and thy Father which seeth in secret himself shall reward thee openly" (Matthew 6:1-4).

We see here that Jesus is encouraging His disciples to give to the poor (alms). Yet, He doesn't do so without showing the right way and right reason to do so while also stating that if a person does alms with the right *how/why* they will be rewarded by the Father. Jesus gave fair value to the people He was leading.

Sell All You Have?

Here's another famous story from the Gospel of Matthew. We'll see again, Jesus always gave fair value when He made any request.

> "Jesus said unto him, If thou wilt be perfect, go and sell that thou hast, and give to the poor, and..." (Matthew 19).

What is the rest of the verse? Take a moment to complete the verse out loud, or write it down.

I have partially quoted this verse and then asked more pastors, teachers, and "Christians" than I can remember to complete this verse. I've never experienced it being done successfully! I have even seen preachers misquote this passage and use it in a way to tell people to just "blindly" or "selflessly" follow Jesus.

Was this your answer?: "Jesus said unto him, If thou wilt be perfect, go and sell that thou hast, and give to the poor, and...come, follow me."

Here's the entire verse (emphasis mine)...

> "Jesus said unto him, If thou wilt be perfect, go and sell that thou hast, and give to the poor, and *thou shalt have treasure in heaven*: and come and follow me" (Matthew 19:21).

Jesus told this man to sell all that he had, give it to the poor, and then he would have treasure in Heaven. This is a spiritual value. This is a reward. Then, separately, Jesus told him to come and follow Him.

Most people omit the treasure in Heaven portion. They present a Jesus who stated that the man ought to give and get nothing in return (one-way justice). They present a "Jesus" who was a beggar. A "Jesus" who was **not** moral. A "Jesus" that didn't give a reason and a value when He made a request.

Actually, Jesus was the perfect leader. He **always** gave a reason and a value when He made a request.

Love Your Enemy

> "But love ye your enemies, and do good, and lend, hoping for nothing again; and your reward shall be great, and ye shall be the children of the Highest: for he is kind unto the unthankful and to the evil" (Luke 6:35).

Notice, even loving your enemy is shown to give you great reward! Jesus only appealed to rational self-interest in order to motivate you to give a value. Jesus is the ultimate leader!

CHAPTER 8

Are, Do, Have

THE LAST CHAPTER ended with the idea that our actions are the result of our morals and ethics. In fact, this is only half of the impact of the law of causality on our lives. Recall that the law of causality says that for every effect, there is a cause.

For instance, I believe happiness and unhappiness are a result of whether people identify things in their lives as a cause or an effect. That is, a cause for why people are unhappy is that they see causes as effects and effects as causes. A cause for why people are happy is that they correctly identify causes and effects.

The most basic and universal example is "Are, Do, Have." As we saw in the last chapter, what we believe determines our actions. Said another way, what we "Are" determines what we "Do." There is, however, a second part to this cause-and-effect process that is equally important—what we "Do" determines what we "Have."

The world tells us the opposite of this. The world tells us what we "Have" will determine what we "Do," and what we "Do" determines who we "Are." Isn't this one of the equations behind advertising? "You need to have this product so you can do this thing and then people (or yourself) will believe you are a certain type of person."

This worldly view is extremely short term because it is trying to take a shortcut by believing that effects prove causes. You see this when people lament their lack of money or possessions because it doesn't allow them to be profitable members of society. For example, I have seen numerous examples of people saying they need to offer the same workers more money so the workers will become better at their job. This is "Have, Do, Are." This statement believes that having will cause people to be better. Their focus is on "Have."

Are, Do, Have

This is not what God tells us to focus on. Think about what God tells us. God tells us to work on ourselves. Yet God tells us over and over again to be profitable. In fact, God tells us He will make us profitable. Is profitability a bad thing? We've already seen that profitability is an objective measure of salvation. In fact, our measure of "good" is profitability. The process is everything.

Next to the kingdom of God, Jesus talked about money more often than any other topic. In all of Jesus' discussions about money, He saw money as an effect, not a cause. Money was the result of doing the right thing. Money is an objective measure in this physical world that cannot be faked in the long term. That is why some people hate money. Profitability is an effect of what we "Do," which is an effect of what we "Are."

Jesus' focus was constantly on the spiritual— on the "Are"— but Jesus related to us on our level by talking about money. All of those discussions were meant to tell us there is an objective measure in the spiritual world that can't be faked, just like money can't be faked in the long term in this physical world. Jesus said the spiritual treasures wouldn't be trusted to those who couldn't handle the less valuable physical treasure: money (Luke 16:11)—not because money in and of itself is important, but because it is an objective effect.

Have, Do, Are

Most people tend to focus on "Have" so they don't have to be profitable any longer. They don't want to be good. Basically, people are hoping they don't have to "Do" anymore. In this instance, "Do" is work, but "Do" is not confined to an occupation.

What people "Are" does come out in what they "Do." If a person commits illegal acts, they are ignoring what they "Are" and trying to justify it by what they "Have." We call these perspectives the "worldly view"—this is a view that values this present physical existence as pre-eminent. This physical perspective emphasizes the "Have" and the "Do." This is not what God wants from us. He wants us to have a spiritual view where the "Are" is preeminent.

Paul said if someone didn't work they shouldn't eat (2 Thessalonians 3:10). Paul also said that we are to work so that we have something to give to those who are in need (Ephesians 4:28). Clearly, working is an important part of our being because it is what we are supposed to "Do." In fact, we are supposed to continue to work and be profitable until the day we leave this existence.

Remember, working is not confined to an occupation. In the parable of the talents, we saw everyone is given a value from God and is supposed to trade with others in order to achieve a profit. You can retire from an occupation, but you can't retire from interacting with people.

Jesus told a parable about a man who had acquired enough that he didn't have to work anymore. The man decided to stop working because he had everything he needed. In the parable, God took the man's soul that very night (Luke 12:13–21).

God wants us to keep working and being profitable. In the parable from Luke 12 referenced above, God took the person's physical life once he expressed his will to stop being profitable. God believed there was no reason for that person to live. God's focus is on the "being." God looks on the inside of the person, and He motivates us by telling us the eventual result of "being" is "Having."

Remember though that effects may not result from just one cause. You can't perfectly predict the "being" of another because of the "Having." People can win the lottery. People can be having a tough stretch that will eventually result in growth. In fact, most often people who "Have" the most materially are not the most profitable according to God. Don't forget—we can have our spiritual value now in a material form or we can leave it up in heaven.

Finally, not all "Having" is material. "Having" also involves emotional, mental, and spiritual value. In fact, a person with a spiritual view realizes these values are superior and cannot be bought with money.

Applications

"Are, Do, Have" is a great measure when talking with people. You can determine a person's depth/emphasis in how they describe themselves and others. The shallowest people describe themselves by what they "Have." "I am rich." "I have a big church." "I'm a millionaire." All of these are "Haves."

Most people aren't this shallow. Actually, the biggest fallacy occurs when they describe themselves by what they "Do." It is very hard in our culture for most of us to separate what a person "Does" from what they "Are." "I am an engineer." "I'm a pastor." "I'm a mother." While we've seen that what a person does is important to God, it is not who the person is.

Look at how God described Himself. When Moses asked God whom should he say sent him, God responded with "I AM." (Exodus 3:14) That is the ultimate "Are"! We saw Jesus described Himself in a similar fashion when He said, "I am the way, the truth, and the life." We can get closer to this perspective when we describe ourselves in terms of our principles. It shows what we value. For example: "I am a person who wants to get closer to God and grow in understanding."

This process also tells a lot about how people see other people. The most instructive example is the definition of homosexuality. I love to ask people, "What is a homosexual?" Most people define homosexuals by what they "Do." Very few describe them by what they "Are." Today, there are TV shows that define homosexuals by what they "Have"—what they wear, where they live, and how they decorate their living space. In fact, the recent "metrosexual" fad is a good example of focusing on the "Have." Basically, "metrosexual" refers to men who share the same "Have" as homosexuals without sharing the "Do."

God focuses on what people "Are." We should do the same. We should look past the "Do" in others and to their "Are" or justice will bring double judgment on us. First, we deserve judgment because we aren't perfect in our "Do." Second, we are hypocritical for holding others to a higher standard than what we are able to attain (Romans 2:1–3).

Summary

As stated earlier, I believe some unhappiness comes from treating causes as effects and effects as causes. We can't completely control the "Do" and "Have." For example, we can't all be athletes or inherit money. We only have complete control over the "Are." That's why God tells us to focus on our inner selves, and then the "Do" and "Have" will naturally become profitable for our uniqueness. We need to work on our being. After all, we are not human havings or human doings. We are human beings.

- How would you describe yourself?
- What did the law focus on?
- What does grace focus on?
- Do you think everyone who "has" the same possessions "does" the same thing?
- Do you think everyone who "does" the same job has the same "are"?

Joel Swokowski's *Commentary*

It's not new information for most Christians that we are each unique. However, outside of this book, I've yet to hear a pastor teach how to determine who you **are** at a level where you can intentionally live out that uniqueness. Yes, we're all children of God. Yes, we are all loved and precious and formed by the hand of the Almighty. But how am I different from all the other children of God? Understanding this "Are" is the key to my own healthy happiness. In fact, people do not suffer from depression when they understand who they "Are" and are making progress in it. Actually, this is key for people overcoming most of the mental health issues in the world today.

Many of the people who attempt suicide do it because they don't know who they "Are." They think what they "Have" or "Do" is who they "Are," and once they realize they are never going to have a mansion or be a rockstar, they don't see any reason to live. If you can't be who you were made to be, there won't be any reason to live.

However, once people understand who they "Are," they realize they can "Do" and "Have" other things and be happy. We will be covering how to determine who you "Are" in two chapters.

Job

Before we move on, let's look at an example from the Bible that shows the "Are", "Do", "Have" Model…

In Job chapter 1 we see that Job was perfect. In the Bible, perfect meant "maximum profitability" (we will see this in two chapters). For now, perfect means there was nothing more (Reward) Job could have done. He feared God (cause) and turned away evil (effect). From the very first sentence, it was established that there was an actual man named Job who lived in a specific location, and he did not do anything evil. This was not a parable.

God asked Satan where he came from. Satan gave God a report that proved that Satan was not omnipresent. Satan can only be in one place at a time. Here is an excerpt from my Bible Commentary blog: The Book of Job. (https://swobible.blogspot.com/2012/01/job.html)

> "And Jehovah said unto Satan, Hast thou considered my servant Job? for there is none like him in the earth, a perfect and an upright man, one that feareth God, and turneth away from evil" (Job 1:8 ASV).

God asked Satan if he had considered Job. God Himself confirmed that Job was perfect, feared God, and turned away evil.

> "Then Satan answered Jehovah, and said, Doth Job fear God for nought?" (Job 1:9 ASV).

Satan's question essentially stated that Job feared God as an effect, not a cause. God stated Job was perfect because he feared God and the effects were that Job turned away from evil. What did Satan believe was the cause of Job fearing God?

> "Hast not thou made a hedge about him, and about his house, and about all that he hath, on every side? thou hast blessed the work of his hands, and his substance is increased in the land. But put forth thy hand now, and touch all that he hath, and he will renounce thee to thy face" (Job 1:10-11 ASV).

Job had a hedge protecting him. This was analogous to the Garden of Eden which also was a "hedge" according to the etymology. Adam was protected by God in the Garden of Eden, but it was his responsibility to "dress and keep it" (Genesis 2:15). In that verse, keep meant "watch, guard, shepherd." Adam didn't do that because the serpent made his way in and led Adam astray. Here, the implication was that Job also had protection from God, however, Job did "keep" his "hedge" and Satan didn't have a way to get to Job. Satan needed "help."

Satan said that if God took away Job's possessions (HAVE) that Job would renounce God to His Face.

God believed Job was good and the effect was that Job had possessions. Satan believed Job had possessions and the effect was that he was good.

In chapter 2, we see again that God asked Satan if he had considered Job. This time, God added more to His description of Job: he still held fast to his integrity although Satan moved Him against Job to destroy him without cause. God was right!

Satan said that if God struck Job's body in order to limit what he could "Do," Job would renounce God to His face.

God believed Job was good and the effect was Job did good things.

Satan believed Job was able to do good things and the effect was Job acted good.

God allowed Satan to do his plan. God put all of Job into Satan's hands except for his life. Life was "the ability to repair." Job still had the ability to repair. Life could not have meant "existence" because there would have been no way for Satan to be right if Job's existence was ended.

Satan struck Job with boils from his feet to his head. These boils would be on the upper layer of Job's skin. This skin was not really "alive." The upper layer of Job's skin would have eventually fallen off. Satan must not have been allowed to affect Job's brain or his internal organs because that would have been a better way to get Job to curse God to His face.

Job's "Do" and "Have" were gone and all that was left was his "Are." For the rest of this story (Job 3-42), we are looking at a man who was solely in his "Are." This is why the Book of Job is a perfect example of the "Are," "Do," "Have" Model. We can see how each of these concepts is separated out in the early parts of the story.

This story also shows God proves His truth through contrastive thinking. God is humble enough to consider He is wrong and was proven right once all the other explanations were proven wrong. Once again, to "consider" I could be wrong does not mean I am wrong and does not mean I have to believe the opposing perspective. It simply means, can I hold that opposing perspective within my thought process at the same time as what I believe? The Book of Job is a great story in that it shows God does have the ability to "consider" He could be wrong, even when He **never** is!

CHAPTER 9

Moral Code

WE HAVE SEEN that what we have is an effect of what we do. We've seen what we do is an effect of what we are. If we are supposed to work on what we are; then what are we? We saw that God is a set of non-contradictory principles and we are made in God's image. Likewise, each of us is a set of principles; however, since we are all unique, we can say that each of us is a unique set of principles. I call this a moral code. We are supposed to work on our moral code.

What is your moral code? It is the prioritization of your principles and shows through in your actions. How can you determine the order of priority? When you are faced with a situation and you have more than one option, the action you choose is a result of the principle that has a higher priority on your moral code. For example, would you rather help a friend or go drinking? In movies, usually, when a character is introduced they will show the character making a choice. This is done to quickly show you the moral code of the character.

Everyone has a unique moral code that is a result of his or her DNA and experiences. I call this the "Physical" or "actual" moral code because it is the code the individual actually uses. Considering the number of principles and the permutations of their prioritization, the individual's

Physical moral code is more unique than fingerprints. If there are 150 principles (money, friends, animals, comfort, growth, etc.), there are 5.7 x 10 to the 262nd power combinations (that's a 5 with 262 zeros after it!). Clearly, this is more unique than a fingerprint.

Because the moral code determines what you do, the moral code is who you are. Also, the moral code eventually results in what you have. Therefore, the moral code can be profitable or unprofitable to the individual. In fact, there must exist a unique moral code for each person that leads the individual to maximum profitability. This is who God intended them to be.

For now, let's call the most profitable one the "Spiritual moral code" and the less profitable one that the individual naturally has the "Physical moral code." The amount of difference between the individual's Physical moral code and his Spiritual moral code varies from person to person. Some are born with less of a difference than others (e.g., Jesus? Mohammed?). Maturity is really the process of modifying one's Physical moral code into his unique Spiritual moral code through reprioritization.

Changing Your Code

Your focus in this world is to convert your Physical moral code over to the one that is most profitable for you. This is harder to do than it seems because society tries to convince you that we all have the same moral code. (Isn't that the point of some commercials?)

Ultimately this is done to control you. Groups of people can't be controlled until their uniqueness is denied. This occurs with organized religions that try to control people by saying they should all have the same "right" moral code. Society tries to control people by saying they should have the same "practical" moral code (e.g., wealth, sex, possessions, etc.).

Converting your Physical moral code to your Spiritual moral code is an act of creation. Recall that creation depends on truth and the individual's ability to detect it. This is going to require thinking. In fact, one could say that thinking is man's only moral responsibility because it is the cause of everything else for the individual. Your ability to live is dependent on your ability to think (create knowledge), and ultimately, morality is choosing to think in order to understand this existence. Conversely, choosing not to think is death. This is done in many fashions: silence, refusing to believe something regardless of the facts, focusing on discrediting or belittling, ignoring information, faking reality, lying, etc. So the first step is to choose to think and understand.

It is going to take energy to convert from your Physical moral code to the Spiritual moral code, so your second step is to identify sources of energy—for now, let's call this "joy." Joy is the result of doing those activities in the moral code that create more energy over the long term. This profitability occurs within the individual. This source of energy is unique to the individual. What brings you joy doesn't necessarily bring me joy. This requires self-awareness.

Take some time to make a list of at least twenty activities that give you energy in the long term and twenty activities that drain your energy in the long term.

We need to be able to look at ourselves honestly and objectively. For instance, one of the things that gives me joy is music. If I have something I don't want to do, all I have to do is play some music and I suddenly have the energy to complete the task. I understand, however, that

I would be mistaken to believe this is true for everyone else or, worse yet, try to make this true for everyone else.

One final note on this step—if the activity gives us energy in the short term but ends up taking energy over the long term through guilt, regret, etc., then this is not joy. If the activity distracts us so we don't have time for self-awareness, then we are just avoiding thought. For example, how much snowmobiling is profitable?

The Survival philosophy would cause us to snowmobile just enough to be able to handle the things in this existence that drain us; however, this is not profitable. The Life philosophy would cause us to snowmobile enough to get energy so that we can invest it in areas that will yield more value. Entertainment can be profitable, but we shouldn't spend so much time snowmobiling that we don't have the time or the energy to examine our lives. This would not be profitable.

So far, the process for converting your moral code is as follows:
1. Choose to think (understand)
2. Identify at least twenty sources of long-term energy

- How would you describe yourself?
- How is this different from your answer at the end of the previous chapter?
- What things do you repeatedly do that are most destructive?
- What did you do on the best day of your life from your childhood?
- Have you made your list of at least twenty things that give you energy and twenty things that drain your energy?

Joel Swokowski's *Commentary*

Another simple way to think of the differing moral codes we carry with us is this:

> **Spiritual moral code:** The exact uniqueness God created you to be in order to achieve your purpose.
> **Physical moral code:** The actual principles you live by.

We hear people talk about how stable a person is and refer to it as "resiliency." The definition of "resiliency" is "the ability to get yourself to a conscious thought process and remain there despite distractions from your surroundings." It involves thinking and being true to your uniqueness rather than being like everyone else. So, the difference between someone's Physical moral code and Spiritual moral code could be seen as a measure of their resiliency. In fact, one of the main things a person progresses in when they grow in resiliency is getting closer to their Spiritual moral code. This also can be seen as transforming your Physical (actual) moral code into your Spiritual moral code.

We all will grow in our uniqueness for eternity. We will all grow in getting closer to our Spiritual moral code. We're flawed, we make mistakes. How we respond to those mistakes is what is important to our growth and to God. A person who is excellent at responding well to their wrongs can be seen as "elastic." Elasticity is the ability to bounce back from losing this conscious thought process.

The exercise of determining twenty drains and twenty gains is intended to help you identify where your energy comes from, and to help you keep it. We are humans, not robots. We run on energy. When we lack energy, it becomes very difficult to maintain a conscious thought process, let alone work on acting according to our Spiritual moral code more.

CHAPTER 10

Determining Your "ARE"

THE THIRD STEP in converting your moral code is to determine your #1 principle. In order to gain control of your life, you must determine the #1 principle on your Physical moral code. Your #1 is the principle that has always won out in every conflict with your other principles. I call this #1 principle the "Physical Are."

Make a list of the decisions you have made during your existence. Identify the choices (e.g., people you dated, jobs you took, places you lived, etc.) and the alternatives. Then identify the principles that caused you to make the choice you made instead of the alternative.

Every decision you make involves a conflict between at least two principles of your moral code. You will go with the decision based on the principle that is highest on your Physical moral code. You may regret it because it shouldn't be the highest, but you'll still go with it until you modify your Physical moral code. That's not to say the #1 was the principle that won out on your major decision—it may not have been one of the choices. The #1 is the one that has never lost.

These principles could be friends, professional power, sex, wealth, and animals—I say the last one because I like to use the PETA people as an example, especially the most extreme ones. When they say that animals shouldn't be used for anything, including medical testing, then their Physical moral code values animal life over human life. In fact, it looks like animals are their Physical Are. Be careful what your Physical Are is because it will lead you to do things you may regret in the long term.

Earlier, I mentioned I use music to give me the energy to handle the tasks I don't enjoy. This technique comes from the Survival philosophy. I am getting energy from one source only to consume it in a way that doesn't give me a return in energy.

Ultimately, your top principles should give you energy. This is the Life philosophy. When your top priorities give you energy, you end up with an abundance of energy and the process becomes easier. Top priorities that consume energy should be prime candidates for substitution.

When we talk about your top priorities on your Spiritual moral code, I call this the "Spiritual ARE" or just "ARE." In fact, it seems most people have two principles that make up their ARE. Like God, these two principles tend to be a qualitative and a quantitative principle. These two principles are who God created you to be for eternity. Your joy can become complete because these are the two principles you will be operating in throughout eternity. The more you operate in them, the more joy you will have. In fact, you can be perfect in your ARE.

The Party

What is the non-contradictory definition of perfect? Perfect means, "to achieve the maximum profitability out of a situation." Every other definition of perfect leads to at least one contradiction with the Bible. Notice, this is not a rigid example that applies to everyone. Perfect depends on both the uniqueness of the individual and the uniqueness of the situation.

The ultimate goal in eternity is for everyone to achieve perfection. This does not mean that everyone is going to Have, Do, and Be the same. This is comparative and results in condemnation. It means everyone is going to eventually get the maximum profitability out of his or her ARE. This is "The Party." How is this possible?

We've seen that Paul compares each of us to a part of the body. When each part does its role perfectly, then the whole body benefits. We are supposed to develop the unique ability we were given by God (ARE) in order to do our job perfectly and fit in perfectly with others who are doing their job. This is the journey we continue after we physically die.

This is God's plan. This is what it is all about. This is why we are here. This is the meaning of life! Your ultimate goal is to find and operate in your ARE. This is your calling.

We are the bride of Christ. Again, we will make up a body. The "cells" of this body are made up of unique beings that have specific talents. Now we can see why God first allowed us to be souls in charge of a physical body, because it is righteous and just for God to let us first be souls that give direction to cells in a body before we take direction from a "soul."

We will want to perfect our ability because it will lead to fullness of joy. Notice how this "meaning of life" takes into account knowing God, having an ultimate relationship with God, connecting with others, eternal joy, loving God, worshiping God, etc. This "meaning of life" model gives the reason "why" we will have the effects espoused by other "meaning of life" models.

From the beginning, I spoke about a party where you could experience any sensual pleasure you wanted to do for as long as you wanted. In order to try to approximate the level of joy you will feel in God's plan, I had to relate it to your current physical existence.

I'm sure the word "sensual" got your attention; however, the key phrase was "for as long as you wanted." There are a lot of things we like doing for minutes or even hours; however, if we were required to do these activities and these alone for days, we would not get the same joy as when we first started.

During the party, the activities you do for eternity will only result in more joy as you become better and focus on them alone. God's plan requires people who desire to operate in their ARE for eternity. The only way this is possible is for each being to receive pleasure from operating in their ARE. In fact, some forms of these pleasures exist for us today.

God made you with unique talents, for a unique purpose. When you do activities today that use some of these talents (ARE), you experience true joy. The amount of joy is dependent on the amount of your ARE that is being used.

I'm not talking about short-term energy bursts that quickly fade or become unprofitable in the long term by turning into guilt. I'm talking about those moments when you do an activity and with every part of your being, you feel your energy increase. In fact, you can look back on it and still get a little burst of pure profitable energy from remembering it.

The Antenna

We saw that God's ARE is righteous and just. We also saw that God spiritually speaks into the individual through grace. If the ARE is the essence of the spiritual person, then God speaks into the individual's ARE. The ARE never changes. It is your uniqueness! The ARE is the antenna that detects God's voice. I look at it this way (Figure 10):

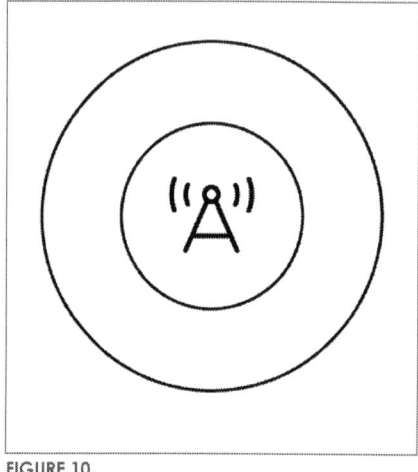

FIGURE 10

The "A" is the ARE through which God speaks to us. In fact, God's grace is on everyone. God is constantly speaking into every person's ARE.

We can be anywhere at any moment in relation to our ARE. The closer we get to the ARE, the more clearly we hear God's voice and the more profitable we are. Outside of the largest circle, we don't hear God's voice and we are not profitable. Just inside the largest circle, we faintly hear God's voice and we are profitable, but in Survival mode. Inside the smallest circle, we consistently hear God and operate in Life mode.

Our ARE can only be perfect when God is flowing through it. This is the only part of our being that God flows through. Everything else is an effect of our ARE. Since God's ARE is righteous and just, that ARE would make every other ARE perfect when God flowed through it.

I like to think of God flowing through our ARE in the same way white light flows through a gem. The light is the source of energy, but the unique colors and patterns we see are the result of the energy flowing through a unique gem. The result is dependent on the uniqueness of the individual, but the source is ultimately from God. Paul said, "But by the grace of God I am what I am: and his grace which was bestowed upon me was not in vain; but I laboured more abundantly than they all: yet not I, but the grace of God which was with me" (1 Corinthians 15:10).

Even today, the ultimate profitability attained by each person occurs through his or her ARE. This is consistent with grace. God is supplying the energy, while the only thing the individual does is just be themselves and let God flow through them. The result is a perfect value that can only be attained through this unique individual, while the cost to the individual was negligible because it comes with little to no effort. Using the profitability equation, there isn't a way to obtain more profitability than getting a perfect value while spending little to no value.

Techniques to Determine Your Uniqueness

The ultimate way to determine your ARE is through grace. Ask God and He will direct your attention. In order to help you begin this process and get a general idea of where to direct your attention, I would ask you to consider your answers to a few questions.

We've already discussed the first two questions: what activities give you energy and what activities drain your energy? This is because when God flows through your ARE, it should lead to a Do that energizes you. The ARE facilitates the unique purpose for which you were created. If the Do doesn't match your ARE, the result is unprofitability, which you will notice as a drain of energy.

If the Do partially matches your ARE, then there will be some profitability. The more the Do matches your ARE, the more profitable you are and the more energy you will have. Notice, there are several possible profitable Do's that result from a specific ARE.

I also asked you to think about what was the best day (or time) of your life. What was a day that caused you to gain energy (even today) when you think about it? Sometimes, it is easier to think back to a day before your teenage years when you felt truly alive and hopeful for the future. The reason you felt energized is that you were using some portion of your ARE. Remember, we'll use our ARE exclusively and our joy will be complete in eternity.

Everyone's ARE is unique. This is much more unique than personality profiles. Whenever you take a personality profile, you are really determining your Physical Are. Besides, these personality profiles are almost "racist." They claim to describe the entire population of the world with (for example) sixteen descriptions. Notice, most people find these physical (Do and Have) profiles helpful for a month or two. Then they move on because they realize two things:

1. They are more unique than the description they read
2. The description doesn't resonate over the long term with the person God made him or her to BE

Some of these profiles justify the Physical Are, preventing the person from operating in their purpose.

The final question that has been presented is: What do you do that is most destructive? Our most destructive actions tend to be the result of a misapplication of our ARE. What makes it most destructive is that we keep doing it. The reason we keep doing it is because it is our

ARE. We are going to keep looking for opportunities to let it come out in what we Do.

For example, someone whose ARE involves "organization" may apply this trait in order to control people and hinder the uniqueness of others. We need organization. We need this person to operate in his or her ARE in order to increase the long-term profitability of everyone else. Unfortunately, this misapplication of their greatest strength can become their greatest weakness.

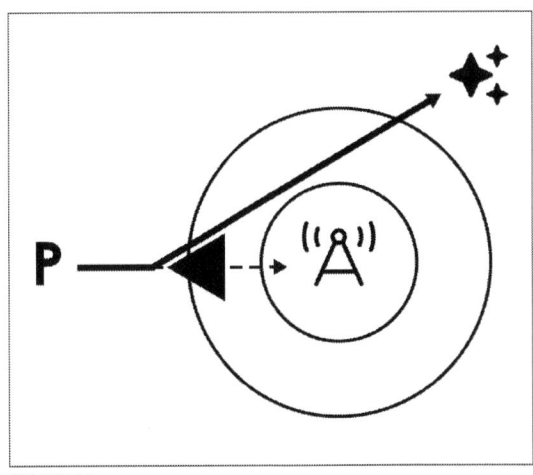
FIGURE 11

The devil is spiritual. He can see your Spiritual ARE. I look at it this way (Figure 11):

In Figure 11, the "P" is where we are when we operate in our Physical Are. We are drawn to our ARE ("A") and want to follow the dotted line; however, the enemy (solid triangle) wants us to keep a Physical focus, so he distracts us. We follow the solid line to the distraction (starburst).

Notice, on the way to the distraction, we pass close enough to our ARE that we can actually hear God and are profitable in the short term. This makes us think we are being led by God. Once we arrive at the distraction, however, we don't hear from God. At this point, we are unprofitable and confused. Even though we know the result was destructive, we will keep repeating the cycle because we are drawn to our ARE. It leads us into our purpose. Operating in our Physical Are creates in the short term, but doesn't create in the long term. Our ARE always creates in the long term.

Converting Your Moral Code
1. Choose to think
2. Identify at least twenty sources of energy
3. Identify the decisions you've made
4. Determine what principle has never lost
5. Operate in grace to determine your ARE
6. Consciously choose to operate in your ARE

Modeling People

If people are "programmed," then the moral code is the "software." Once you figure out what a person's hierarchy is in their values, you can predict their actions. Be careful whom you share your Physical Are with; don't just blurt it out to a group. In fact, understanding the Physical moral code of another can be used for evil, by allowing you to control him. How can you determine the Physical and Spiritual moral code of others?

In addition to the questions that have been posed, the most immediate way is to observe the choices a person makes. Since your understanding of another person's Physical moral code is limited to the number of choices observed, there is only so much that can be learned from observation alone. Also, this is not completely accurate, because we are acting as if effects come from only one cause. It is going to take a lot of observation to remove all the other possible causes to the effects we see. Since our accuracy increases with the number of observations, the best way to intentionally learn more is to pose conflicts to the person being observed.

For example, the first sentence of the first book is, "Imagine you are invited to a party where you will be able to participate in any sensual pleasure you desire for as long as you want." The phrase "sensual pleasure" makes some people uncomfortable because they think it means "sexual immorality." They see this sentence in conflict with a book on God.

Sensual pleasures are any enjoyment that involves the senses. This could be eating a great meal, listening to music, or hugging your kids. Since the sentence leaves it up to the hearer to choose ("you desire"), respondents are actually telling us about themselves when they respond to the conflict posed in this sentence.

Jesus was a master at taking actions different from His intentions in order to determine the moral code of others. When the poor woman asked for a healing, Jesus acted like He wasn't going to heal her in order to hear her responses and determine her moral code (Matthew 15:22–28).

After Jesus rose, He appeared to His disciples and at one point made as if He was going to venture on (Luke 24:28) in order to determine the moral code of the disciples. Jesus was not being dishonest. Jesus was presenting conflicts in order to observe the individual's choices. This also explains why we learn more about people when they suffer as opposed to when they triumph. Suffering causes us to make hard choices.

The Physical Are can appear to be a strength but always ends up looking weak when examined by others during times when there isn't a crisis.

For instance, someone's Physical Are may be "progress." He will always make the choice that allows him to progress immediately. Usually, when this is examined more closely, it appears this person isn't patient. His decisions aren't righteous because he doesn't have enough of the right data or he isn't interpreting the information objectively. This Physical Are ultimately results in unprofitability in the long term.

One final thought on diagnosing the Physical Are is to realize the Physical Are is the individual's attempt to compensate for a less-than-perfect God and the destructive things the individual does. Every person gets their impression of God from their father. Since no father is always completely right and fair, we all start out with the impression that God is less than perfect.

Ask someone to describe his father. This is the easiest way to determine someone's perspective of God. If he says his father was never around, then he won't believe God is around. If his father was abusive, he will think God is abusive. If his father was lax with discipline, he thinks God is easygoing. Since everything begins with God, the first book began with helping the individual receive a non-contradictory understanding of God.

The enemy wants to keep the individual from attaining his ARE and hearing from God. As the individual grows used to not attaining his ARE and being destructive, he begins to believe he isn't valuable and his life has no purpose. This also leads the individual to believe God must be less than perfect because He created a being without a purpose. It looks like the enemy has won, and he has, as long as the individual doesn't think.

The individual then creates a law unto himself to compensate for his destructive behavior and his perspective that God is less than perfect.

This is the Physical Are. This will also change over time depending on the individual's experiences and circumstances.

Beliefs are unconscious and uncontrollable. Thoughts are conscious and controllable. As long as the person doesn't know his Physical Are, he is not able to intentionally change his behavior. The process described in these books has helped numerous people determine their Physical Are and Spiritual ARE in order to make intentional changes in their lives.

If you understand the moral codes of others, you can predict their actions; this leads to control. After all, what is persuasion if it isn't understanding your customer well enough to offer him something he values more (is higher on his moral code) than his current belief? When you understand a person's Physical moral code, you can control him—you can speed him up and you can bring him to a halt. You can help him make or avoid unprofitable decisions.

On the other hand, when you understand a person's Spiritual moral code, all you can do is get value from them when they choose to operate in it. In fact, you can get a perfect value that can't be found anywhere else. One of my personal goals is to help every person I interact with find their ARE and help them operate in it. If I am also in my ARE, then every interaction will be profitable for everyone involved.

Not only should we want to understand our ARE, but we should also want others to understand it as well. It is the key to profitability and can never result in destruction. Recall that God isn't defensive about acting in His ARE. He is righteous and just. Every interaction with God will only result in God being right and fair. Recall also the big difference between God and us is that God can't act outside of His ARE, while we have the ability to act apart from our nature (Physical Are) and intentionally choose to operate in our ARE.

Summary

In the first book, we discussed growth vs. comfort and contradictions. People tend to let contradictions exist in areas they are comfortable in. These areas tend to be high on a person's Physical moral code; otherwise, they would eliminate the contradictions.

Understanding what is #1 on your moral code (Physical Are) allows you to regain control over your decisions. We can take control and change conscious thoughts. We can't take control or change unconscious beliefs. Identifying your Physical Are explains the why behind the decisions you've made and points you towards long-term profitability. When people identify their Physical Are, they instantly understand the why behind the decisions they have made.

The Physical Are is a result of your experiences and circumstances, so it is going to change over time. The ARE never changes because it is who God made you to be for eternity. Understanding your ARE leads to long-term profitability and stability. Remember, our ultimate goal in eternity is to be connected to others, so we are **all** made with this need to connect. This is not our ARE because it isn't unique to the individual.

Again, the ARE is unique. It would take another book to walk you through all the steps to intentionally find your ARE. We have a website that lists the latest techniques, testimonials, and examples of how others found their Physical Are and ARE. In fact, there is a quiz according to Romans 12 that will help you get a big-picture direction for determining your uniqueness. For further information go to: www.modelinggod.com.

The purpose of understanding all of this is to show us what to do to increase the profitability of our interactions. The key to understanding how is love.

We now have everything we need in order to understand love and its implications.

- How does understanding your Physical Are explain the decisions you've made?
- How can understanding your Physical Are help you during times of crisis?
- What are things you can do to get into your ARE?
- What are things others can do to help you get into your ARE?
- How does understanding the ARE of others help you have profitable interactions?

Joel Swokowski's *Commentary*

This chapter is what sold the original publisher on this book. He recognized that "determining your ARE" was something no one else was doing! This also tends to be the information that even those against Lenhart and this book stay away from. In fact, over the years of helping people determine their ARE, even those people who have estranged themselves and slandered the author have still used this information to better their lives!

The author called personality tests "almost racist" because they attempt to put the entire world's population into 4 to 16 categories. This is a long chapter and Lenhart may not have felt the need at the time to go into more detail. Here is more explanation about personality tests:

Every personality test comes from a theory created thousands of years ago called, "The Four Temperaments." The creator of that theory (Hippocrates) stated that whenever he saw four specific behaviors come out of people he knew they were unbalanced! They had too much or too little blood, bile, black bile, or phlegm.

Notice, personality tests are based on what a person looks like when they are unbalanced. This means personality tests aren't a measure of a person's Spiritual ARE. Instead, they help a person get a big picture direction for their Physical Are!

Is this a surprise? The word personality comes from the Latin word *persona*, which means "mask." A personality test measures the "mask" a person hides behind when they are unbalanced. Again, another description of the Physical Are.

The personality of a person is *what* we see, and the ARE of a person is their causes, their *how/why*. We see an application of this in the Bible when Samuel goes to Bethlehem to find the next king of Israel. He sees the sons of Jesse and wrongly chooses when he's focused on the appearance (*what*), not the causes (*how/why*).

1 Samuel 16:7 says, "But the Lord said unto Samuel, Look not on his countenance, or on the height of his stature; because I have refused him: for the Lord seeth not as man seeth; for man looketh on the outward appearance, but the Lord looketh on the heart."

One of the first lessons I learned on my journey to resiliency is that I am *not* my behavior, I am *not* my brain, I am *not* my mistakes, and I am *not* my outward appearance. This is good news! This means I'm not defined by the mistakes I've made. The uniqueness I am defined by (my soul) is the intangible being within me that God created, and that BEing is able to be discovered and intentionally manifested. That is what this chapter is facilitating in you!

Check modelinggod.com for a quiz that will give you a big-picture direction for you to determine your uniqueness. Also, there are resources to help you determine your purpose, which is how your uniqueness is profitable in a specific context.

Remember, the author's goal was to help the reader get a unique explanation for who God created you to be so you can achieve spiritual fulfillment. A personality test actually prevents your ability to do this because it is focused on the opposite of your Spiritual ARE.

CHAPTER 11

Love

THE MOST POPULAR passage concerning love is 1 Corinthians 13:1–8. The King James Version of the Bible uses the word "charity" in place of love:

> "Though I speak with the tongues of men and of angels, and have not charity, I am become as sounding brass, or a tinkling cymbal. And though I have the gift of prophecy, and understand all mysteries, and all knowledge; and though I have all faith, so that I could remove mountains, and have not charity, I am nothing. And though I bestow all my goods to feed the poor, and though I give my body to be burned, and have not charity, it profiteth me nothing. Charity suffereth long, and is kind; Charity envieth not; Charity vaunteth not itself, is not puffed up, Doth not behave itself unseemly, seeketh not her own, is not easily provoked, thinketh no evil; Rejoiceth not in iniquity, but rejoiceth in the truth; Beareth all things, believeth all things, hopeth all things, endureth all things. Charity never faileth: but whether there be prophecies, they shall fail; whether there be tongues, they shall cease; whether there be knowledge, it shall vanish away" (1 Corinthians 13:1–8).

Note the emphasis on profit in the first three verses. Again, you are required to be profitable. Everything you do is measured by its ability to yield a profit. According to this passage, love is the key to profitability. In fact, you can't be profitable without love!

The remaining verses describe love by listing the effects of love. It is a list made up of sixteen results. Nine of these are actually results applying to what love is not (e.g. "envieth not"). This passage doesn't tell you what love is; that is, the cause of love.

This passage sounds like the Henderson 4000. There is no description of the causes of love in this passage. How do you intentionally have more or less love without knowing the causes? This is knowledge, not understanding. We'll have to look elsewhere for the definition.

Exercise

In fact, take a moment to answer the following questions. First, what is your non-contradictory definition of love? Write it down. We tend to fool ourselves by thinking or speaking a definition, but it can't really be tested, understood, or examined until it is written.

Next, what do you mean when you say, "I love you"? Again, write it down. Is it consistent with your non-contradictory definition of "love"? Please take a moment to at least attempt to answer these two questions.

The Bible says in Romans 13:8–10, "Owe no man any thing, but to love one another: for he that loveth another hath fulfilled the law. For this, Thou shalt not commit adultery, Thou shalt not kill, Thou shalt not steal, Thou shalt not bear false witness, Thou shalt not covet; and if there be any other commandment, it is briefly comprehended in this saying, namely, Thou shalt love thy neighbor as thyself. Love worketh no ill to his neighbor: therefore love is the fulfilling of the law."

Here, love is an effect. In fact, it looks to be the ultimate effect; it is the fulfillment of the law. Recall in the first book, works were an effect of grace and faith. Yet, people are so focused on being right, they try to shortcut faith and grace by making themselves and others focus on works. They are focused on, "Have, Do, Are," instead of, "Are, Do, Have." They do the "right" things for the "wrong" reasons.

People make the same mistake when it comes to love. Instead of realizing we all are given righteousness and justice from God and have to intentionally choose to work towards love, people try to shortcut the process and have love. People don't know the causes. The proof is the lack of people who can write down a non-contradictory definition of love or the meaning of the phrase "I love you."

Love is the ultimate effect and in order to intentionally have it, we need to understand the causes of love.

Definition

In the Bible passage, the word "charity" is used. Today, most people define charity as "giving without getting anything in return." In a concordance, the word is agape. Its meaning is "love in a benevolent way." The root of this word is, "love in a social and moral sense." Love involves giving to others, not getting. For the moment, let's combine the definitions and define love as "the giving of value to another person without getting anything in return from that person."

We saw there are three ways to exchange value with others. The exchange of value involves both giving and getting. If love is focused on the giving portion of the exchange, why should you give without getting anything in return from that person?

In other passages, the Bible says you should give without expecting anything in return from the debtor. In fact, Luke 6:33–35 equates giving value and not expecting a return with love, saying,

"And if ye do good to them which do good to you, what thank have ye? For sinners also do even the same. And if ye lend to them of whom ye hope to receive, what thank have ye? For sinners also lend to sinners, to receive as much again. But love ye your enemies, and do good, and lend, hoping for nothing again; and your reward shall be great, and ye shall be the children of the Highest: for he is kind unto the unthankful and to the evil."

You now know the reason you should do this comes from your understanding of prayer and justice. If you give and expect a return from the person, then you will continually look to them for the return. When the return doesn't come as quickly as you like, you may worry and may even resent the person to whom you gave. You can even end up owing (Figure 7, Panel 4). I think we all can see this isn't love.

Let's be clear. There is nothing wrong in and of itself with giving and expecting a return from the person; it's just that this action is not "love."

Actually, you should expect a return; however, your focus should turn to God. If you don't expect a return, it is like prayer when you turn it over to God. God can give you a value in the spiritual realm (Figure 8, Panel 2) because you have given up worrying about what you are owed. God can invoke justice because He is assured you won't seek your own justice. After all, Jesus said, "Give, and it shall be given unto you" (Luke 6:38).

Now you know how to intentionally increase in love—you need to give without expecting anything in return from your debtors. You also know why you are supposed to do this—you love so that you can become more profitable. These causes agree with the description of the effects of love at the beginning of this chapter; however, there is another reason why you would give without expecting anything in return.

Four Pillars

In order to understand the other reason for love, you need to understand how four complex concepts are connected: morality, uniqueness, justice, and profitability.

Morality — Morals drive (cause) the behavior that results in long-term profitability (goal) for the individual. Morals are true for all places at any time in history.

Uniqueness — We are all different. We value things differently. In fact, we value the form of the value we get differently. One person may prefer money. Another person may prefer a gift. Still another person may prefer public recognition. If we were all the same and valued everything the same, we couldn't be profitable in our interactions.

Justice — Everyone gets exactly what he or she is owed. You can't get a value by only giving something less. You will eventually have to make up the difference.

Profitability — Our goal is to increase our value over time. This comes from uniqueness and justice! God built profitability into this world when He created uniqueness. When I give you something in exchange for something I value more and the thing I gave is more valuable to you than what you gave to me, then we are both profitable. The cause of profitability is God (justice).

The only other reason you would not expect a return is because you are the debtor and you are trying to get back to even. Love is also the giving because you believe you owe. Love begins when you realize that someone has given you more value than you believe you would have given to get that value.

If someone has quantifiably given you more value than you believe you've given to them in order to deserve it, then you owe him or her. If you are honest (Spiritual moral code), you will repay it by giving the necessary quantity of value back to that person until you believe you're guiltless. If you are dishonest, you won't give it, but deep down you will be guilty in your own eyes.

The result of profitable morality is more love; that is, you are focused on giving value whether you feel like it or not. Look at the passage again. This result is what is stated. So the key to intentionally loving is to be moral and aware of justice—recognize the value others have given you and pay them back. Uniqueness will multiply the value the receivers get in their eyes. If they follow their Spiritual moral code, they will pay you back and the process can continue for eternity.

Love is an action (result) based on (caused by) the individual's morality. It is not a feeling or emotion. Also, since these exchanges involve justice and uniqueness, God still plays a role in the exchange of value between individuals. Ultimately, God is the cause of profitability between people.

- What is an example of love you think you have been giving?
- How should profitability factor into the exchange?
- How should justice factor into the exchange?
- How should uniqueness factor into the exchange?
- How should morality factor into the exchange?

Joel Swokowski's *Commentary*

A pastor once said to me, "Wait a minute Joel, love has multiple definitions. I could say, 'I love my tacos' and I could say, 'I love my wife,' and these are two completely different meanings."

I've had this conversation before, and it sounds plausible and logical. Yet, when I pushed this pastor on his definitions, even the "multiple" he claimed to have, he wouldn't answer me.

We humans have turned the word "love" into a concept that can mean multiple things. There are between 4-7 different Greek words for love, depending on the "scholar" you choose to reference. The fact it's not even agreed upon how many definitions there are for this one word shows the flaw in this argument. However, the main point is that there are multiple different Greek words that **we English-speaking people** have translated into **one** word: Love. The flaw in the English language compared to the Greek used in the original scriptures is not a defense for having multiple definitions for one word. This is a translation error.

Furthermore, although the pastor definitely meant different definitions of his love for tacos and his love for his wife, he did have **one specific definition** for each usage of the word love. He didn't mean two things each time he used the word love, he meant one thing each time he used the word even if that meaning was different each time.

There is one definition for love that Jesus gave to us and that is the standard that Christians are meant to live by when following the greatest of the commandments: from the Greek word agape: "To give a value without expecting anything in return from the person to whom you gave."

This is God's definition of love; this is Jesus' definition of love.

I've had more issues with the definition for love with pastors and church leaders than anyone else. In fact, the people who don't attend church, and even atheists tend to be appreciative of the definition taught in this book. I think God created us to have an instinctual understanding of His definition for love. Love must be objective, and it must not **only** be a feeling. Yet, the difficult part of this definition for everyone, whether you agree with it or not, comes down to "without expecting anything in return."

Expectations

The expectation part is the big piece the church is missing. People give, but people usually expect something back from who they give to. This chapter does help us understand that I can expect something back from God when I give to another person. Having faith and understanding in God's Justice and Reward will help you love in a way that makes it easier to not expect back from that person: you'll know that God will pay you back in the long term.

Jesus said there is no greater love than giving your life for a friend. Notice, you are giving the greatest amount and you can't expect anything back, because you are dead. In fact, look at how the "effects of love" (charity when grammatically necessary) passage begins:

"Though I speak with the tongues of men and of angels, and have not charity, I am become as sounding brass, or a tinkling cymbal. And though I have the gift of prophecy, and understand all mysteries, and all knowledge; and though I have all faith, so that I could remove mountains, and have not charity, I am nothing. And though I bestow all my goods to feed the poor, and though I give my body to be burned, and have not charity, it profiteth me nothing. Charity suffereth long, and is kind; charity envieth not; charity vaunteth not itself, is not puffed up" (1 Corinthians 13:1-4).

Do you know the root cause of anger? Expectations. If I have an expectation that is not met, it results in anger. **Love** is the key to dissolving my anger problems! Every time someone is angry, all you have to do is ask, "What were you expecting?" Not having this definition of love, God's definition, is enabling anger issues!

Notice what else is shown in these verses: Paul stated you could give without love. This means that love is both the giving and the not expecting. It's a conjunctive. I need both halves of this definition for it to be true, for it to be love: "to give a value (freedom) without expecting anything in return from the person to whom you gave (limitation)."

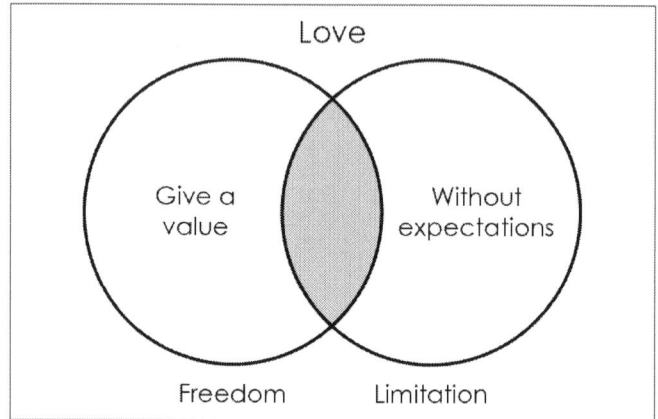

CONJUNCTIVE ILLUSTRATION FOR LOVE

330

MODELING GOD — THE MEANING OF LIFE

Jesus' Definition

Luke 6 shows Jesus being contrastive. He covers the two ways to give that don't result in love with the only profitable option being "giving without expecting anything in return."

> "And if ye do good to them which do good to you, what thank have ye?" (Luke 6:33).

In other words, if I know ahead of time that they're going to pay me back, what "good" is that? What does that create?

In the next verse, he continues, "for sinners also do even the same. And if ye lend to them of whom ye hope to receive, what thank have ye?" (Luke 6:33-34).

The word "hope" means "to fully expect"... not to love! Jesus continues:

> "For sinners also lend to sinners, to receive as much again. But love ye your enemies, and do good, and lend, hoping for nothing again" (Luke 6:34-35).

A great example of love is when you give to your enemies because you know you won't get anything back from them. Finally, let's look at the last part of this passage:

> "And your reward shall be great, and ye shall be the children of the Highest: for he is kind unto the unthankful and to the evil" (Luke 6:35).

Four Pillars

This is arguably the most important principle for a person to understand and live out as it relates to "Christian Living." We are known by our love for one another. The greatest command comes down to loving God and loving your neighbor. So why did Lenhart wait so long to get to this doctrine? I mean, the book's almost done!

These "Four Pillars" explain why the journey to love took so long. We need to understand Morality, Uniqueness, Justice, and Profitability in order to understand Love.

Morality is the cause, it is expressed through our uniqueness and justice ensures the result is profitability.

CHAPTER 12

Love Applications

THE REASON LOVE is so important is that it is the end (result) of the law. The Bible says, "Now the end of the commandment is charity out of a pure heart, and of a good conscience, and of faith unfeigned" (1 Timothy 1:5).

The goal of the commandment is love, lack of guilt, and solid faith. Now we understand how they are connected.

We are created in God's image. We all have a sense of righteousness and justice naturally; no one has to learn the desire to be right or to be upset when someone is unjust to them. God is love because He followed righteousness and justice to their logical fulfillment. In order for us to follow God, we have to choose to love. It has to be intentional because it doesn't happen naturally.

This definition of love also lends itself to the idea that there are greater and lesser loves because a person can give greater or lesser value. The Bible says the ultimate value is one's life.

> "Greater love hath no man than this, that a man lay down his life for his friends" (John 15:13).

Notice there is no way for the person who gives the value (their life) to get a value in return.

Saying, "I love you" is not the same as giving a value. If saying, "I love you" is the declaration that a value has been given, this boast or show of pride isn't love. When a moral person says, "I love you," they are saying they recognize they owe a value. This is actually a humbling experience. This admission of being in debt makes the person vulnerable. This admission is made unilaterally and does not require anyone to say, "I love you" back.

In reality, this is not what most people mean. Otherwise, how can they get so offended at you so soon after saying they love you? This person experiences vulnerability because he or she wants you to respond in kind. When most people say, "I love you," they are speaking of an emotion. This love is known as eros.

The cause of confusion with the term love comes from society's primary use of the word to describe an emotion. You can't control an emotion to the point of intentionally feeling a specific way all of the time. This love cannot be intentional.

It seems like everyone wants to be loved; however, if people love me it is either because:

1. They know they owe me and are trying to get back to even, or
2. They are focused on giving in order to build long-term value through justice.

Either way, I gain a short-term value when people love me and lose a long-term value. People who are focused on being loved are really hoping for a short-term value and little to no long-term value. People who are focused on loving others are giving up a short-term value in hopes of receiving a long-term value. Jesus focused on loving others and didn't get offended when people didn't love Him, because it was their loss and His gain.

You are supposed to love people and use things. This means that you are supposed to give a moral value to people and get a value from things. Many people get this quite confused when they focus on getting a value from people and give their time, energy, and focus to things. Remember, love is social and moral. It requires interacting with individuals and is focused on profitability. Every interaction results in an exchange of value.

Jesus asked Peter if he loved Him. When Peter said he did, Jesus responded by saying, "Feed my sheep" (John 21:15–17). Notice, Jesus was showing Peter that love is a cause that results in an action, just like faith, grace, etc. We are seeing a pattern here.

Acting in Love

When we are confronted with unjust behavior, the Reward Model says the best way to receive a long-term value is to forgive and forget. The quicker you forgive and move on, the greater the spiritual value you will retain. Does this mean we should never confront unjust behavior? Actually, we should, but only if we can do it in love.

There are two ways we can confront unjust behavior. The most common way is to retaliate. Even if what we are saying is a fact, if we do it in order to feel better about ourselves, then we are gaining the value back. Any action that results in us receiving an immediate short-term value is really some form of retaliation.

The less common way is to do it in love. This means we confront the unjust behavior with the attitude we are trying to help the people who are behaving unjustly. In essence, we are giving them a value. If they don't want to hear it or agree with us, then we respond according to the Reward Model: we quickly forgive and move on with our lives without resentment towards the unjust people.

If our attitude was truly "in love," we would not care whether we got a value back from the person we were helping. If we can't approach this person with this heart attitude, or more likely, if we can't walk away from this person after they've refused to hear us, then it is best in the long term for us to avoid the confrontation and follow the Reward Model.

Furthermore, we are supposed to speak the truth in love (Ephesians 4:15). Recall, earlier in the book, we looked at how speaking the truth is more demanding than speaking facts. Now we see the Bible makes a further demand. How do we speak the truth "in love"?

One example: a congregation member could tell their pastor that people think another preacher is a better speaker. This may be a fact. It is only a truth if it can be used to create something of value in the long term. Even if this were the truth, how would it be spoken "in love"?

If the member told the pastor with an attitude he was giving the pastor a value, then this is "in love." If it was spoken in order to make the member feel better or to make the pastor feel bad, it was not spoken "in love."

Summary

Love is the ultimate effect of righteousness and justice. Love is the giving of a value to another person without expecting or requiring that person to return a value. The Bible consistently speaks about doing everything in love. Our actions should be focused on what we can give instead of what we can get. This action can be done intentionally, regardless of the individual's feelings, because justice takes care of any inequalities.

Love is also righteous because it is the only right way to interact with others. This is the ultimate example of how we are supposed to interact. Any interaction apart from love is only going to result in guilt or the potential to achieve less than the maximum profitability God had intended for us.

We have seen that God is righteous and just; however, these principles result in two separate models that guide our day-to-day living. The Salvation Model relies heavily on righteousness. The Reward Model relies heavily on justice. The goal of both of these models is profitability, which is dependent on the universal principle of uniqueness.

Looking back to the party analogy, we could say that profitability occurs as long as we stay between Righteous Street and Justice Avenue. We know that righteousness and justice converge at God, but how far out do they radiate from God?

While the Salvation Model and the Reward Model become separate, they both rely on uniqueness and ultimately converge again in love. The ultimate way to get reward is love. Since reward is based on justice, the only way to receive the maximum reward is to give to people and don't expect anything back.

Likewise, if you follow God's grace completely, it will result in love. God will constantly tell you who to give a value to and not to expect that person to return a value. In fact, because God initiates the exchange, His intention is for us to use the value to create the maximum possible profit. This creation of value results in us getting the maximum just portion of this created value if we don't expect anything in return because justice will ensure we receive half of the created value. Expecting a return opens us up to the possibility we will spend some of the value by violating justice.

Furthermore, while following grace in love also results in maximum profitability via the Reward Model, because God initiates this exchange, it also results in salvation. Salvation depends on being righteous and not sinning. The ultimate way to avoid sin is to give to people God has chosen and not expect them to return a value. There is no possible way to sin if this is done. This explains the how. Next, let's summarize what we need to do to be profitable.

There are three levels of creating value. The first level is through justice. While this creates a value for us, it is not due to profitability. The second way is through profitability—exchanging value with people who value things differently than us. This higher method is still governed by justice. The ultimate method is grace—God giving us a value and instructions on how to use it—God flowing through our ARE and us choosing to let it come out. Not only does this result in the maximum reward if we do it in love, but it also results in salvation.

Everything Jesus said concerned either the Salvation Model or the Reward Model. He either spoke of following grace to get the same ultimate value (salvation) as everyone else or doing works to use justice to accumulate more or less reward than everyone else. Most books focus on one model and ignore the scriptures describing the other.

Jesus summed it up when He said, "Thou shalt love the Lord thy God with all thy heart, and with all thy soul, and with all thy mind. This is the first and great commandment. And the second is like unto it, Thou shalt love thy neighbor as thyself. On these two commandments hang all the law and the prophets" (Matthew 22:37–40).

The first commandment covers the Salvation Model. The second covers the Reward Model. Notice, both are made perfect in love.

Giving a value and not expecting a return from that person goes against our nature. The only way we can act in love is to act apart from our nature. This only comes from God and occurs through grace. Grace and love are the *"what"* and the *"how"* to achieve perfection.

We have now completed our model of God. We have derived all of God's qualities from righteousness and justice. We have united the models in the ultimate Godly characteristic that results from righteousness and justice: love. This is why the Bible says, "God is love" (1 John 4:8).

Since we understand profitability, we can see that love can be infinite. There is no limit to love; it can go on eternally especially if unique beings are able to exchange value in their ARE. Going back to our party analogy, Righteous Street and Justice Avenue radiate from God until they meet in love, and love goes to infinity.

Now we know why we are supposed to love. Next, we will look at the most visible example of love.

- What do you mean when you say, "I love you"?
- What value was attached to the last time someone said, "I love you" to you?
- How would you correct someone by doing it in love?
- How would you explain to someone that love is the ultimate effect of righteousness?
- How would you explain to someone that love is the ultimate effect of justice?

Joel Swokowski's *Commentary*

We saw in this chapter that there are two reasons a person is motivated to love:

1. Embracing justice and reward in hopes of a greater reward in the long term.
2. Owing the other person.

The first reason is what all Christians ought to strive to grow towards and grow in. Loving a person because it's the right thing to do and doing that more often and in more situations is one of the ways I can grow in spiritual maturity.

The second reason is what a moral person does. This person recognizes that justice is real and that it would be a benefit to give fair value back to the person who gave it to you. This would also describe a person of integrity. I know that even when I'm not expecting anything in return from the people I give to, when they give back I appreciate it. A person of integrity treats others the way they want to be treated, and if I want people to give back to me, then I ought to give back to the people I owe. This is where we are in our relationship with God **all the time**, I'll never give back to Him enough where I don't owe Him!

One of the things that helps people gain the meaning of "I love you" is to spend a month saying, "I owe you" instead. Then after a month, saying, "I love you" will have the intended meaning to both people.

In 1 Timothy 6:10, the phrase "love of money" is not love (agape). It is the only usage of the word and it means "friend of silver." The implication is that in this case, money is seen as a cause and not an effect. This also serves as a great example of what I explained in the previous chapter.

This is more accurately viewed as a translation issue. One English word was used for two different Greek words. Does this give me an excuse to rationalize that love can't be defined?

God is Love

In I John 4, "God is love" is used at the beginning and end of a passage as a form of proof. The passage implies that only God is truly able to love, and if we love others it proves that God is doing it through us.

In this epistle, the Apostle John modeled love:
> We ought to love one another because love is of God, and loving is done with understanding. Loving shows you know God. Everyone that loves is born of God and knows God. This means you can only love (as a cause) if you know God and if God (Holy Spirit) is in you.

Model Summary:
> When you know God, you will love, because God is love. This does not mean God always loves. It meant that only God loves.

This model was analogous to the critical verse, 1 John 1:5, which states that "God is light, and in Him there is no darkness at all." That verse was used to prove whether a person was a believer or not by whether they had light because the light can only be obtained from God via the Holy Spirit. Likewise, if you do not love, you do not know God, because God (alone) is love.

Critical Point:
> The correct interpretation of "God is love," which occurs twice in 1 John 4 because it is the beginning and proof of a logical

argument, is "as to God's Nature, it is loving." It is an effect of the cause (God's identity/nature).

God's Friend or Enemy?

Notice, God first loved us. Since He doesn't expect anything in return, our response shows if we are a friend or an enemy. A friend would give back, which results in salvation. An enemy would take the value and then act like they weren't loved. People who say they would believe in God if He gave them something are overlooking what God originally gave to them and making themselves God's enemy. God doesn't predetermine who is His enemy. He loves everyone, and people use their free will to determine if they are God's friend or enemy.

When Love Becomes Enabling

It's important to understand this definition of love, otherwise a person can enable someone and think they are loving them. If the value a person gives begins to cause the one who receives the value to become destructive, continuing to give to that person would be enabling them to be more destructive.

The "value" from love is meant to be a value for the person I'm giving to, not necessarily to me, the giver. If I give $20 to someone dealing with drug problems knowing that the person is going to use the money for drugs, the result is damaging in the long term and is therefore enabling a destructive path. This is true regardless of how much value I see in the $20. If the value I give is being used for long-term destruction, it is objectively not a value and not love. In fact, this could be seen as a form of deception: right *what* ($20) with a wrong *why/how* (destructive use), which proves that enabling is destructive over the long term.

God realized He would have crossed this line from love to enabling if He had continued to give to Israel. That is why He told Jeremiah He was done giving. It is why He spoke of profitability. It is why He said not to pray to Him because He wouldn't hear.

God's Nature

God's love is guided by righteousness and justice. This is how we know that love is **not** God's nature. The contrast of this is the answer to this question, "What guides God's righteousness and justice?" **Nothing**. Those of the causeless causes: the **First** Cause.

CHAPTER 13

Marriage

WHILE SOME CONSIDER marriage the ultimate result of love, I believe it is the ultimate test of love. Over time, a marriage will expose whether the participants love in a social and moral way, or not. What is marriage?

Most people see it as a contract. A contract assumes failure and is based on distrust. It only is useful when one of the parties fails to uphold their responsibilities. In fact, one of the definitions of the word contract is "to make small." Clearly, it is a very limiting word.

Marriage is a covenant. A covenant is an agreement where two parties share all that they have, all that they do, and all that they are with each other. It is based on trust. They give each other complete access to everything they "Have, Do, and Are." They do this to cover their weaknesses. They offer their strengths to the other person in order to get the other's strengths in place of their weaknesses. Uniqueness and profitability enter in again. It isn't an agreement you want to enter with someone who is the same as you. Where is the profitability?

The Bible actually goes one step further and states the result of the covenant is the individuals share one flesh. They are supposed to merge the physical in order to focus on the mental and spiritual aspects. Keeping with the theme of this book, marriage is the ultimate example of individuals exchanging value with each other. In fact, the Bible says the marriage covenant is a three-way covenant—the two individuals and God. The individuals are supposed to completely share their physical attributes with each other and each is supposed to share their spiritual attributes with God.

Today, people see marriage as a contract that can be broken at any time for any reason. If the marriage was formed without God, then all that needs to be considered is what the two people think because they were the only two participants in the marriage. If the marriage was a covenant formed with God and all of the participants have a say, then God needs to be consulted.

God says a covenant can only be broken by death. That's why the Bible says the requirements of marriage no longer apply to widows. We have seen, however, that life and death apply to more than just the physical. The spiritual life and death of the individual are actually more important to God. The Bible does allow for the ending of the covenant in cases of the spiritual death of one of the participants. We will look at this in detail in a subsequent book.

Emotion

Today, marriages have such a poor success rate. There are two main reasons. We've addressed the first one, which is that people don't understand marriage is a covenant, not a contract. The second one was mentioned in the last chapter—people think love is an emotion.

At first, when people are in love, one of the results is a very intense emotion. This burst of energy is wonderful and serves as the fuel that progresses the relationship in the short term. The person in love doesn't think about his or her long-term needs; whether it is food or sleep. They are focused on giving because the value they receive in the short term is overwhelming, but they can't sustain this feeling over the long term. They will need to eat. They will need to sleep. At some point, the emotion is supposed to lessen just enough for the person to intentionally create a relationship that will continue to be profitable in the long term. These are not the people we are talking about.

We are talking about the people who see love as an intense emotion. These people want to focus on sustaining this intense emotion over the long term. They equate this intense emotion with the value of the individual and get married while this emotion is still the focus. Consequently, they expect this emotion to never lessen.

These people think they are marrying the perfect person for them and are surprised when one "more perfect" comes along or their mate is no longer perfect. Their emotions change and they conclude they are "out of love" with their spouse or have fallen in love with a person outside their covenant. I like to illustrate these changing expectations with a theory I call "The Concentric Circle Theory."

The Concentric Circle Theory

When you date, you are determining how far your date is from your ideal person. Imagine the ideal being a point (the star in Figure 12). The person you are dating (A) is some distance from this point. In fact, this person describes a circle around your ideal point (Panel 1). When you date another person, preferably you would date someone closer to your ideal.

Said another way, this person would be within this circle. This new person (B) describes another circle concentric within the first (Panel 2). Realize you may not actually start out with an ideal and can be learning what your ideal is by dating. If your ideal changes, then the process begins anew. Each time the person is closer to the ideal, they will seem more perfect and you will have more intense emotions.

If there is only one person for you, then you will keep dating until you find the point or the star. We know that marriage is a choice and there is actually more than one person you can marry. The people you would create a covenant with make up an inner circle (Panel 3, shaded). The goal of dating is to continue the process until you find someone within the acceptable inner circle.

Your acceptable circle can change. Some people start out with one that is very large and then are disappointed when they marry someone far from their ideal. Some people start out with one that is small but over time they increase the size of the circle.

The marriage will fail if you focus on trying to find someone closer to your ideal, instead of focusing on the person you have married. Your ideal will change over time, even after

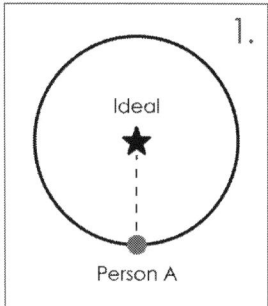
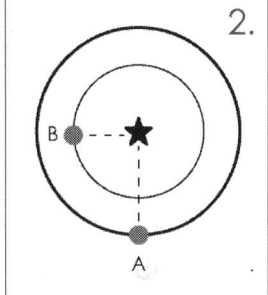
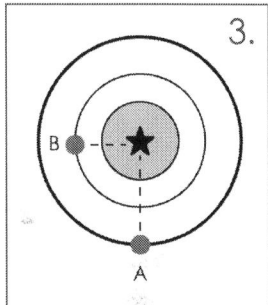
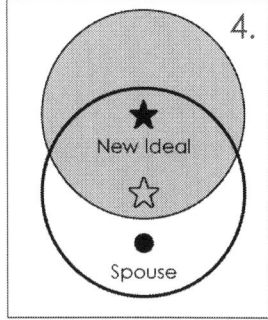

FIGURE 12

you get married. The goal is to find someone near enough to the ideal that adjustments to your ideal won't leave them outside the acceptable circle. In this way, another possible cause of a failed marriage is that your mate (M) may no longer be within your acceptable circle (Panel 4).

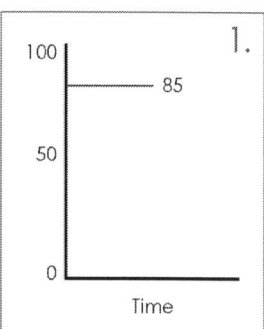

Change in Love

This theory is a simplistic explanation to help people set expectations prior to marriage. When we look at couples that are already married, the reasons marriages fail are better explained by looking at love. The health of a marriage can be measured by love. Most failed marriages could have been saved if the individuals understood how to measure the health of a marriage and intentionally fix it. The key is the change in love.

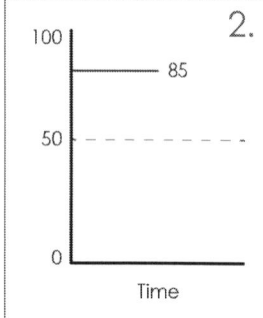

Most people judge the health of their marriage by the "absolute value" of love. It's easy to see now why this happens. Let's say a person is getting value (love) from their mate and their marriage is at 85 (Figure 13, Panel 1). They may be happy as long as the marriage doesn't fall below 50 (Dotted line, Panel 2).

As the amount of value (love) exchanged diminishes, the rating decreases. As long as

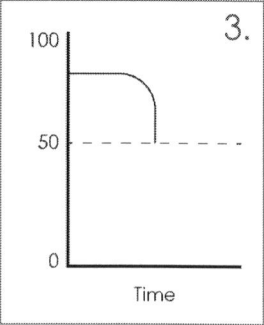

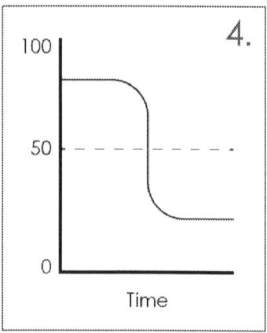

FIGURE 13

the rating is above 50, the individual isn't going to take any measures to increase love (Panel 3). Because love involves an exchange, a lessening of love accelerates the rate of decline.

Eventually, one person is going to stop giving because they believe they are owed a value.

If the person doesn't do anything until the rating hits 50, then it will be too late because by the time he or she addresses it and slows down the slide, the marriage may be at 10 and either party may feel it's too far gone (Panel 4).

Furthermore, some people stay in the marriage even though it is down to a 7. It's at this point that one of the individuals says the marriage isn't great but at least their spouse takes care of the kids or puts a roof over their heads. In this example, that value is worth a 7, and the individual has adjusted their expectations from a 50 to a 7.

We have been speaking, however, as if there is only one measure of love for their marriage. Actually, each person has a measure of the love in the marriage. The key is to be aware of both your measure and the measure of your spouse. Otherwise, you may be getting a value, yet your spouse may be running out of love.

Physical Are vs. ARE

If both people are initially focused on their Physical moral code, it is easy to see why most marriages aren't successful. When the covenant is formed, it is between two people who believe the interaction of their current Physical Ares is profitable.

We have seen that Physical Ares can't be perfect and their profitability is limited, but the most important point is that Physical Ares change. They are a result of the experiences and circumstances of the individual. This change in the Physical Are is either going to lead to the marriage getting stronger or it getting weaker (Figure 12, Panel 4).

If the couple had understood each other's ARE, this would look totally different. It is possible to have a profitable interaction with everyone as long as everyone is in his or her ARE. First, once the couple learns each other's ARE, they can figure out how to profitably interact with each other. Second, the couple could determine the opportunities where the two AREs overlap.

Actually, this applies to all our interactions. When we interact with some people, we may have to narrow the scope of our interactions. For others, we may find we can be completely in our ARE while they are completely in their ARE. This latter situation is ideal for marriage.

The ultimate way to improve the profitability of our interactions is to understand our ARE and the ARE of others. This is especially critical when it comes to marriage.

Summary

Marriage is a covenant relationship. It is the ultimate example of two people exchanging value because each person vows to share all they "Have, Do, and Are" with their partner. It is something that can be done intentionally because the vow is dependent on an action, not a feeling. If the vow taken before God is dependent on a feeling, then it will eventually fail. No one can promise to feel a certain way indefinitely.

Love is the measure of the strength of the marriage. If the individuals consider love to be an emotion, then the marriage is eventually doomed. You can't intentionally feel the same way indefinitely. When the couple made their vows, those vows needed to be over something they had control over. You can intentionally increase the love in the marriage if it is a choice to give a value. More to the point, because love is a choice, you are able to still love your spouse even when you don't feel like it.

Finally, the marriage that is formed over the Physical Are's has about a 50% chance of failure because the Physical Are is going to change over time and the profitability will either increase or decrease.

- Do you know any marriages that have endured over the long term based on emotion?
- What are the benefits of a covenant relationship?
- Where do you get your examples of love and marriage?
- Thinking of a loving relationship you know of, how has it changed in the last five years?
- What are three ways you can love someone when you don't feel like it?

Joel Swokowski's *Commentary*

Trust

This chapter begins by emphasizing the truth that marriage is based on "trust." The word "trust" means the same thing as "faith": a belief in something you cannot see, either because it is invisible or has yet to happen. The usages of "trust" in the Bible happen when grammatically necessary. Similar to love and charity: The command says to "love God," not "charity God" due to the grammatical usage of the word. With trust and faith you would say: "I trust God", not "I faith God".

Trust and faith being synonymous implies that trust is also built through understanding and experience.

Experience v. Understanding

Most marriages fail and it is because we don't have a tool to determine the ARE of each person to see how much they are going to have to adjust. The author laid out the experiential issues while stating the previous chapters in the book would have explained the understanding.

Marriages Gone Wrong

Similar to Lenhart, I am a fan of CS Lewis. One thing I learned from Lewis has impacted the way I see love and whether or not it's an emotion. In this chapter, we saw one area where marriages can go wrong is when people believe that love is an emotion. How many marriages begin with the traditional vow to love your spouse from this day 'til your last, 'til death do you part (or some variation)? Whether or not this vow is even possible, it exposes a contradiction in the idea that love is an emotion. What I learned from Lewis is that love must be something I choose to do for this vow to work. If love were an emotion, and I promised to love you for any length of time, it would be no different than me promising to never have a headache again. There are things I can do to prevent a headache, and there are things I can do (causes) to help myself feel a certain emotion (effect), but I will never have direct and complete control over how I feel.

There is a reference to another book that will come and bring more detail to marriage, divorce, etc. Without getting into much detail now, I do want to help you understand the end of a marriage covenant enough to at least help you not get into trouble. (Again, the next book will give a full explanation.) The church has caused much damage around the topic of divorce. I'm not going to spend the time unpacking all of this now. What I want to encourage you to do is simply try to remove the bias we have towards divorce being a sin or wrong by being contrastive on the man-made tradition in this area.

I've heard and read time and time again that "God hates divorce." Is this true? Or is this a translation error? For now, know this, there are two processes that end a covenant, including a marriage: Divorce and Putting Away.

Divorce and Putting Away are two different concepts.

Before a person judges that divorce is wrong and that divorced people have sinned, I'd encourage them to check the root words from Matthew chapters 5 and 19, and Mark 10 that deal with marriage, divorce, and putting away. I'd also encourage them to read Deuteronomy 24 and Isaiah 50.

Energy in Marriage

There are two types of marriage:

1. The spouses get energy inside the marriage and invest that energy outside the marriage.
2. The spouses get energy outside in an attempt to prop up the inside of the marriage.

The first type of marriage is what the church needs from its husbands and wives. These are marriages that, due to the energy they get from each other, are able to help and love and invest into the lives of the people in their community. This is one of the ultimate forms of profitability.

The second group is sapping the resources of their community, including the church. Do I need to go out and play golf with my friends just to make it through another night with my wife and kids?

The reality is, whether you agree with or like the concept of divorce or not, it happens and it happens regularly. This chapter also diagnosed some major issues as to why marriages go wrong and even why marriages begin wrong.

One of the ways I've learned to look at people entering into a new marriage is through a computer hardware and software analogy. How often does a man or woman end a marriage and then start a new marriage thinking, "now with this person, it'll be great!"? Yet, if I had a marriage that ended due to unprofitability and I don't change anything about myself and my approach to marriage, it'd be like me buying a brand new computer but putting in an old operating system. Why do I think a computer built this year would run great using a Windows 95 operating system?

The real issue with my marriage is the software. How am I approaching my spouse? How am I loving my spouse? How well do we work together?

If I don't know how to be a good spouse, it doesn't matter who I marry, it's going to go bad!

The next chapter looks at ways to drive the value you get within the marriage so a person doesn't have to go outside the marriage to get energy to prop up their marriage.

CHAPTER 14

Marriage Applications

THE BEGINNING OF the end for a failed marriage is one or both of the parties believing they are owed a value. They stop focusing on giving and start focusing on getting. Since love is related to giving, the love runs down. For example, instead of giving a value to the other because he is moral and knows he owes his spouse, the husband starts feeling like he is all paid up. In fact, he starts to believe he is owed. This is not love.

There are three causes to the feeling of being owed.

1. **One person forgets the value they've received.**
   Since the process depends on a person recognizing what he owes, the quickest way for the process to run down is for one person to forget the value he's been given in the past. This value can be a one-time exchange or something that happens on a continuous basis. In reality, each person is constantly giving the other a value if in no other way than he or she is choosing to do things for the other person instead of doing his or her own thing.

One common cause is that one person takes the other for granted. They become so accustomed to the values being given to them; they no longer see these values as special. So, he or she no longer believes he or she owes for what was done in the past. Another common cause is that one person decides they don't want to owe any more. For any number of reasons, the person focuses on his short-term gain and convinces himself he isn't in debt; this requires the person to ignore justice. This is usually done by ignoring morals and focusing on ethics.

2. One person's values change.
Looking at the process, another key attribute is the ability of one action to have different value to each individual. This is because profitability depends on uniqueness. Therefore, the process can run down when one person changes the value he credits for the actions that are done for him. We saw this can be a result of a change in the person's Physical Are.

Each of us has tasks we hate to do. These can be simple tasks, but our uniqueness makes these tasks large energy drains, especially for the value we get out of them. The roots of this are in our moral codes. For example, a husband may hate taking out the trash for whatever reason; however, to his wife, this is a relatively minor task. If the wife does this task, she considers the value she's giving relatively small, but to the husband, this is a huge value. Rather than look at these irrational requests as silly, they are actually a huge opportunity for moral couples to become profitable. How can this situation deteriorate?

First, the husband can decide that it is nothing for her to do this task and change the exchange rate of this task. Instead of it being a huge value, he can begin to count it as a small value. His rationale may be that it doesn't cost her very much to do this task. See how he has to justify the change? He has to focus on what it costs her rather than the value he's receiving. This denies uniqueness. This is not love.

Another way this can run down is for her to do the task and berate him over his irrationality. For him, the exchange rate of this task has changed. Now the value he is receiving isn't huge. It is the huge value minus the value she is taking by belittling him. She thinks she is giving a huge value and doesn't realize the value has changed. When he begins to give less value back, she's going to feel she is owed. This cause relies on denying uniqueness and morality. Again, this is not love.

3. One person had the wrong values from the beginning.
This love process is relying on both people actually having a profitable relationship. In order for this to be true, both people need to value uniqueness, justice, and morality. If either of the individuals values something else, the process is flawed from the beginning. How could they miss this?

The dating process is supposed to be used to determine if this is true, but often, the beginning of the relationship is filled with intense emotional feelings. If they get married before they have a chance to work on these issues, the couple may be in for an unpleasant surprise.

Another way for the relationship to be flawed from the beginning is for at least one of the people to value unprofitable attributes (Physical Are) in their mate. We saw with the concentric circle theory that a person identifies an ideal whether he realizes it or not and values the mate in relation to the ideal. If he valued the wrong things at the beginning, the relationship will become unprofitable. Additionally, as he grows, he is going to realize his mistake and change what he values. As he values different things, the value of his mate diminishes.

Again, if a person believes that marriage and love are based on a feeling, then he can find an easy reason to end it. If the person realizes he made a commitment and the marriage is a choice, then he will work to stay married. The fact that marriage is a choice, however, brings up what would seem to be an obvious result. If you don't want to be married, you won't be, because ultimately, marriage is a choice. Additionally, if your mate doesn't want to be married, you won't be married.

At this point, we are not concerned with whether it is "right" or "wrong" for a spouse to change their ideal or exchange rate. We are focusing on why people want to end a marriage. We will deal with when it is "right" or "wrong" to end a marriage in a subsequent book.

Communication and Intimacy

As for working on a marriage, other books could give an exhaustive list of things to do. I would like to briefly address two key areas.

First, the ability to communicate in a fashion that reaches your mate is critical. Since we are unique, we have different ways of communicating. You can intentionally learn the best way to communicate with your spouse. You can intentionally increase and improve your communication with your spouse in order to avoid the three pitfalls listed above.

Communication:

- allows the two people to keep track of changing values and conversion rates.
- helps the transition as someone grows and values things differently.
- addresses a change in currency exchange before someone has liquidated his account.
- identifies an account dwindling before it reaches the point of no return.

The second area is intimacy. The couple needs to open up to each other and become vulnerable. They need to continue to share more over time. This can be done intentionally; however, it can only be done in a loving environment. As long as intimacy is increasing, the couple's relationship is growing.

The process is analogous to what we learned about faith leading to progress. With increased intimacy comes more understanding of and experience with the covenant partner. Understanding and experience are the causes of faith. Also, growth is possible because these shared areas provide more opportunities for the couple to connect.

Here, as in the salvation process, the emphasis is on growth, not speed. This is a marathon, not a sprint. If the couple focuses on intimacy, they will share more over time. This will take communication, which was covered above.

The growth in intimacy will give the couple a deep spiritual value they won't find in anyone else. This will be a value that is hard to liquidate. Again, it requires the couple to become vulnerable. Notice that only in a covenant relationship can individuals feel secure enough to become completely vulnerable.

This brings us back to the theme of exchanging value. After all, the marriage covenant is the ultimate way humans exchange value with each other in this existence.

In eternity, the party will consist of beings in covenant with several other beings through their ARE. This will require the ultimate in intimacy and the ultimate in communication. Every being will know what they are valued for and how much they value every other unique individual. There won't be jealousy, because everyone will be perfect in his or her uniqueness.

Physical beings should serve as a constant reminder of the meaning of life. Every time you look at another person, you are looking at a collection of unique cells working together to accomplish something greater than what could be accomplished separately.

God has created every person with a desire to connect to others through uniqueness. We see this today when people join gangs, fall in love, form cliques, or strive socially to be in the inner circle. God's eternal plan is dependent on connection. The ultimate connection is a covenant.

Marriage is important to God because it is the physical representation of the meaning of life. It is God's way of letting people experience a tangible form of this ultimate connection. Now you can see why marriages are under attack. The last thing the enemy wants is marriages operating according to Godly principles. The enemy wants to create doubt that God's eternal plan is possible and/or desirable.

There are people who want to believe we can't or shouldn't be able to understand God's eternal plan until we get to heaven. That wouldn't be holy.

It is right and fair of God to give us tangible examples of His eternal plan.

- What is an example of someone forgetting the value they received from you?
- What is an example of someone changing the way they valued something you did for them?
- What is an example of someone having the wrong values?
- How do you adjust your communication style to fit the uniqueness of others?
- How can you intentionally improve your vulnerability in sensitive areas to become more intimate?

Joel Swokowski's *Commentary*

From the previous chapter, we now see how crucial it is for people to have a tool to determine their ARE! Here are two measures every marriage ought to consider:

1. How much does each spouse act in their ARE?
2. How much does one of the spouses have to act outside their ARE, so they do not lose the energy needed to keep the other in their ARE?

What is Marriage Really about?

Marriage is about accruing spiritual value, benefiting the church, and profitability for the entire community. We see this is largely missing in the world today. One of the ways I can see this is that people are also largely unaware of what the word "husband" means.

The term "husband" has turned into a word that simply means "the male in a marriage relationship." Its actual meaning is "worker of the land." It is a husband's job to help his wife become the person that God created her to be. The husbandry, or "wife," is meant to be the source of profitability in this marriage covenant.

This also helps us understand why it's beneficial for a wife to submit to her husband. It would be for the wife's benefit! If a husband ever tells their wife to blindly do what he is saying, using the "submit" instruction as his support, he shows he is not loving his wife. Anytime the topic of "submitting" comes up in a marriage, the wife has every right to ask her husband, "how is this for my benefit?" The husband is not meant to benefit from the submission. The husband is meant to facilitate the purpose and progress of his wife.

When a husband and husbandry work in the way God intended, in the way God has towards His children, the result is profitability for everyone involved. How many marriages do you know that fit this description?

Communication

This chapter showed us that we need to communicate to the uniqueness of the individual we're interacting with, especially in marriage. While we all should strive to find a specific way to communicate effectively with the specific unique people in our lives, this will be covered in detail in the next book. For now, here's a quick presentation that will help you communicate more lovingly with anyone:

"Let no corrupt word proceed out of your mouth, but what is good for necessary edification, that it may impart grace to the hearers" (Ephesians 4:29).

To grow and help others grow, edifying communication is critical. This includes: *what*, *why*, and *how* we communicate.

In the Bible, edifying people followed three guidelines in communication:

- Make statements on yourself (Share).
- Ask questions of others (Learn).
- Answer others' questions of you (Share).

The opposite interaction is destructive communication:

- Make statements on others (Accuse/Judge).
- Don't ask questions of others (Pride—Assumption that you know what they are thinking).
- Don't answer the questions of others (Isolate).

Communication, edifying or destructive, is a cause. A "cause" is simply "the start, or how something begins." There are four types of communication causes you can give to one another. Here they are, in order according to the amount of increasing control being exerted upon the other person:

Good Cause: Open-ended questions or statements of fact.
"How are you doing?" or "That wall is green."

Not Bad Cause: Close-ended question or statement of your opinion.
"Did you like that restaurant?" or "I don't like the color of that wall."

Bad Cause: Projection (telling someone what they think or feel) or judgment.

"You don't like that restaurant" or "You're stupid for liking the color of that wall!"

Worst Cause: Negate another.

"What is your favorite color?" They answer "red". You state: "No it's not, it's green!"

Communication Example

Look at the following examples of the three guidelines to see the type of communication (edifying or destructive) that occurred:

> "And they heard the voice of the Lord God walking in the garden in the cool of the day: and Adam and his wife hid themselves from the presence of the Lord God amongst the trees of the garden. And the Lord God called unto Adam, and said unto him, Where art thou? And he said, I heard thy voice in the garden, and I was afraid, because I was naked; and I hid myself. And he said, Who told thee that thou wast naked? Hast thou eaten of the tree, whereof I commanded thee that thou shouldest not eat? And the man said, The woman whom thou gavest to be with me, she gave me of the tree, and I did eat. And the Lord God said unto the woman, What is this that thou hast done? And the woman said, The serpent beguiled me, and I did eat" (Genesis 3:8-13).

God: Asked open and close-ended questions (good/not bad).
Adam: Shared (good); made statements on Eve and God (bad); didn't answer questions (bad).
Eve: Made a statement on the serpent (bad).

If God knows all things, why did He ask Adam questions? It was to get Adam to think, and to give him an opportunity to confess and repent! God gave up control. Adam condemned himself and all mankind with his answer—justifying himself and blaming Eve and God!

There are examples like this all over the scriptures. If you want to stretch yourself, take a look at 1 Samuel 17. The story of David and Goliath starts with some great examples of both edifying communication and destructive communication. Are you able to identify it? Who do you think is the most destructive? Goliath? Or someone else?

Check out the website for more resources to help with marriage: www.modelinggod.com

CHAPTER 15

Living the Life Philosophy

THIS BOOK COVERED two more directions we are given to help us get closer to God—pray and love. Now we understand why to pray; it is the only way to exchange value with God. We also understand why we are supposed to love; it is the only righteous way to exchange value with others. The ultimate example in this existence is marriage.

We also covered two philosophies for how we approach every opportunity to exchange value. The Survival philosophy looks at opportunities to exchange value as limited to a finite amount for a specific time. The Life philosophy looks at opportunities to exchange value as unlimited: we should be giving the maximum value all of the time.

Thinking back to the first book, we can see many applications. The most powerful examples relate to grace and forgiveness.

Grace and Forgiveness

The Survival philosophy looks at grace as something we obtain for a specific time. We run our lives and determine our own actions. When we have hard times, grace is something that can get us out of the situation. We can go to God and let Him direct our lives until we get out of trouble, then we run our lives again. The result is not profitability.

The Life philosophy looks at grace as something we do all the time. We let God direct all our actions all the time. The result is profitability and abundance.

As for forgiveness, earlier in this book, we saw that one of the ways to intentionally get a value from God is to forgive a past injustice. The Survival philosophy would cause us to wait until we needed the value before we would forgive the specific injustice. The Life philosophy would cause us to instantly forgive everyone of every injustice against us. Forgiveness would be unlimited. There is no good reason to retaliate on our own behalf. This never creates.

Getting Out of Survival Mode

When people live according to the Survival philosophy, we say they are in survival mode. The results of living in survival mode for an extended period of time are burnout, depression, anxiety, illness, etc. We need to help others intentionally get out of survival mode. In order to help others, we need to understand two concepts.

First, we have to reconcile the conflict between the Physical and the Spiritual. We are Physical beings. We are also Spiritual beings. What role does thinking play?

The overwhelming majority of people speak as if thinking and the Spiritual are mutually exclusive. Most of the time, this belief is blatant, but there are subtler versions of this belief. For example, we saw in the first book that some people say, "I'm more artistic and you are more scientific." This comment shows the person doesn't think it is possible to be both completely emotional and completely logical. As stated from the beginning of the first book, I believe the Bible says thinking and the Spiritual are not mutually exclusive.

The way to resolve the relationship is to view growth as a progression from the Physical to the Spiritual through thinking. Thinking is a process. In keeping with causality, we would say, "Thinking is a cause, and growth in the Spiritual is the effect." Let's define each area:

> "Physical" — tangible things ("Have") and feelings without a known cause.

> "Spiritual" — intangible things ("Are") and feelings with a known cause.

Hurdles vs. Drivers

The second concept is important for assigning value to different activities. Everything can be categorized as a hurdle or a driver. A hurdle is a required level that must be met in order to have progress, but the profitability in that aspect drops off after that level is exceeded. A driver continues to give a return as the quantity of the aspect increases.

For example, the number one attribute in an automobile that correlates to customer satisfaction is the cup holder. If the vehicle doesn't have a cup holder or the cup holder isn't satisfactory, people see the entire vehicle as unsatisfactory, but if the vehicle has twenty cup holders, the owner does not prefer the vehicle twice as much as one that has ten cup holders.

The cup holder is a hurdle. There is a minimum requirement in this area and continuing to invest in it does not increase the customer's satisfaction. An example of a driver may be gas mileage—the greater the gas mileage, the more satisfied the customer will be.

This understanding of hurdles and drivers is crucial when putting together a financial budget. If a person treats a hurdle like a driver, he will go broke very quickly. The answer, however, is not to follow the Survival philosophy and spend the minimum in every area. The way to increase profitability according to the Life philosophy is to increase your resources in drivers.

Governments tend to run deficits because they treat hurdles like drivers and drivers like hurdles. It is hard to get some people to realize they are putting too much money into a hurdle. One way to help them towards this realization is to ask them if it would be more of a benefit if they bought the gold-plated version. If they say yes, the attribute is a driver. If they say, "We don't need that much," then you have shown them there is an upper limit. From there you can continue to bring them back to a reasonable investment level.

Physical vs. Spiritual

The Physical is necessary and cannot be ignored, but it should be valued as a hurdle. The Physical is bound by time. The Physical will never be perfect. In fact, the Physical is always running down.

The Spiritual is the emphasis and should be viewed as a driver. The Spiritual is not bound by time. The Spiritual can be improved. In fact, the Spiritual can become more perfect.

Throughout history, every Great Mind agrees that focusing on the Physical only leads to destruction in the long term. Yet, we are surrounded by people and information that try to convince us the Physical is a driver. They believe that the Physical is the cause and happiness is the effect. We saw this when we identified this worldly view as believing in "Have, Do, Are."

The Great Minds (Philosophy, Figure 14) knew we had to progress out of emphasis of the Physical; however, their arguments focused on whether we should progress towards logic (Thinking) or feelings (Spiritual). Thinking is a process that leads to the emphasis of the Spiritual.

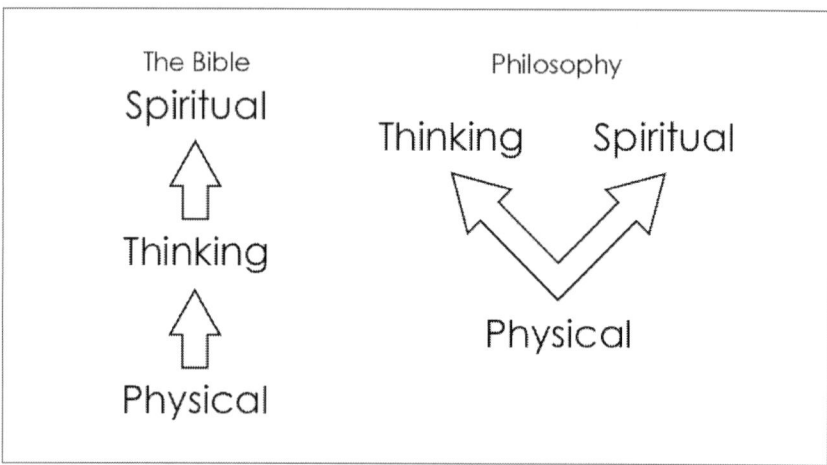

FIGURE 14

People who believe that logic and feelings are mutually exclusive are referring to feelings without a known cause. The Bible teaches us we should have a good attitude and then gives the reason why. The Bible values feelings, yet they are feelings with a known cause. Feelings of this type (emotions) are an effect. This is a Spiritual focus.

This does not mean that the Physical is unimportant. Remember, the Physical is a hurdle. There is a minimum amount of value you have to achieve in this area. For instance, you have to eat and be healthy. If you don't eat or if you become unhealthy, it will be very difficult to be able to focus on driving the Spiritual. Likewise, if you eat too much or focus too much on becoming healthy, you won't get the return you would if you focused on the Spiritual.

Thinking (understanding) is the cause for progressing from a Physical focus to a Spiritual focus. That is why thinking is our only moral responsibility—everything else is an effect.

Physical Model	Spiritual Model
1. Tangible things ("Have")	1. Intangible things ("Are")
2. Feelings are the cause	2. Feelings (effect) from a known cause
3. Hurdle	3. Driver
4. Bound by time	4. Will exist forever
5. Always running down	5. Can be perfected
6. Occurs naturally without thought	6. Requires thought

Application

People operating in the Life philosophy are becoming more profitable over time because they are treating the Spiritual as a driver and the Physical as a hurdle. They use the minimum amount of resources (energy and time) in the Physical and put the maximum amount of resources in the Spiritual. Remember the lists you made of things that give you energy and drain your energy in the long term? These activities are in line with your ARE and are Spiritual.

People in life mode are essentially investing today's energy in activities that give a greater return today and in the future. How are they making it through the day? They are getting energy today from investments they made in the past.

People who are in survival mode are doing activities today that give them just enough energy to make it through today. This isn't necessarily bad. There are instances where we will be in survival mode. The goal, though, should be to end the day at the same or a better level than we started the day. If we are ending the day worse, we are in a survival mode that will not be easy to escape and eventually lead to burnout, destruction, etc. People can see this from the outside.

These extreme cases happen when people do things to get through the day with no thought given to the future. They are essentially borrowing from the future to consume it today. This leaves them in a bigger deficit tomorrow. How do we help others break out of this cycle?

The key is to get them to look at the big picture and the long term. They need to invest the minimum in the hurdles (the things that help them make it through the day) while investing the maximum in drivers (areas that will yield a higher return). What are the drivers?

The drivers are the areas associated with the person's ARE. This is why it is vital to know yourself and who God created you to be. As people increase their resources in drivers, they will eventually see a greater return and have days where they are no worse off. This takes faith because it is something they can't see and that hasn't happened yet.

As they continue to invest more into drivers, they will be more profitable and exist in life mode. Again, this is something people can see from the outside.

Summary

The Bible tells us to pursue the Life philosophy. It tells us to turn everything over to God all of the time. This is the abundant life. The people who actually live this philosophy are easy to identify. They value thinking. They don't overvalue the Physical. They treat the Do and Have as effects.

These people operate in their ARE. They live by grace. One of the goals of these books is to point the individual back to taking direction from God in everything they do. The voices of the enemy and our flesh are loud. God's voice is soft. In order to hear His voice more clearly, we need to choose to ignore the flesh and focus on hearing from God. When we do, we are in our ARE and life mode.

The second goal of these books is to make the Bible more understandable so people will want to read it more often. Understanding the non-contradictory definitions of key biblical words leads to increased understanding. Reading the Bible builds faith. We need to match faith with grace in order to be saved and exist in life mode. These books direct everyone back to God and the Bible through understanding. Ultimately, this is the only way to make progress.

The first book covered intangible concepts: God and salvation. It is nearly impossible to tell who actually believes in God and is truly doing the salvation process because the action is mostly internal. Unfortunately, most people focus on knowing the salvation concepts instead of actually doing the process. The first book focused on helping people understand these concepts so they can focus on doing the process.

This book focused on practical applications. We have covered how we are supposed to exchange value during the journey. It is something that is external and obvious to everyone. These are the areas where the world can point out our contradictions. After all, Jesus says the world will know us by our love (John 13:35).

The only way to truly experience the concepts that apply to our day-to-day Christian lives is to completely take direction from God. If we follow grace in everything, the Bible doesn't promise us happiness during our time on earth. It promises profitability.

Ultimately, the Bible promises eternal life. We saw that eternal life is the ability to eternally repair ourselves. It is also the opportunity to eventually operate perfectly in our ARE and experience complete joy. On earth, however, we are only promised growth and abundant life. This makes sense because life is necessary to make growth profitable.

Growth is painful. It begins with God breaking down some part of us. This is not comfortable. Our response to God's pruning determines our Christian life. If we follow the Survival philosophy by using the ability to repair in a fashion that only mends the broken area, then we are not profitable. God wants us to let Him do more. If we follow the Life philosophy, we will continue to let Christ's ability to repair work in our lives beyond simply mending the affected area and produce unlimited growth. This takes thought (understanding) and the result is profitability. This is abundant Christian living according to Jesus.

- What are three hurdles you have treated as drivers?
- What are three drivers you have treated as hurdles?
- What activities do you do when you are in survival mode?
- What feelings do you have for which you don't know the cause?
- What would your life look like if you were living in life mode?

Joel Swokowski's Commentary

Hundreds of years ago, a book was published that contained the foundational physics principles that are taught today as Laws. It began with definitions and applied them in order to build models.

It barely sold any copies and quickly went out of print. One story is that a student of Newton remarked when Newton walked by, "There goes a man who wrote a book that neither he nor others understand."

However, ten years later, it seemed that all the scientific questions could be answered by the "Principia." The king, when he realized how many different topics were explained by this one book, exclaimed "Is he even human?"

People begged Newton to write another book, but he refused. It wasn't until 1713, over 25 years later, that he published the second of the three books that would become one of the most important works in the history of science.

Likewise, "Modeling God" was published in 2007. This one book contained the only non-contradictory explanation for the foundational Christian doctrines: the models for God and Salvation.

Besides barely selling any copies and quickly going out of print, the book also brought the wrath of religious authorities who went out of their way to personally attack the author. Not only did these religious authorities' actions oppose the words of Jesus, but they also broke their state's law with regard to defamation and committed abominations to the Lord.

The author's willingness to follow Jesus' teaching to not defend himself even resulted in these religious authorities feeling justified in their judgment against him and his book. They actively went out of their way to defame the author even 15 years later.

During that time, pastors were unwilling to meet with the author. Worse, these pastors prevented people from getting help, while not providing effective help to these same people.

In the 15 years since this book was published, not only has it been effective in helping people get off of antidepressants, anxiety, and bipolar medication, but it also predicted the decline of the church in America. If God's Nature is love, and grace is "unmerited favor," then no one is headed towards eternal damnation, and there's no reason for anyone to change his or her behavior.

Ultimately, this book facilitates a person to be restored, as the apostle Paul encouraged Christians to be in Galatians 6. This book helps people intentionally gain control of their lives with what Paul teaches as the ability to bear your own burden. Once a person has that ability to bear their own burden and to have control of their lives, they can then give up control of their lives to God. I cannot give something to someone, even God, that I do not have myself.

In addition to understanding the Bible better, a major benefit readers have obtained from this book is the ability to hear from God more clearly and more often. God speaks truth (doctrine) through grace. Our ability to understand what He is saying depends on whether we have the same definitions for the words God is using. This book helps with the key words (God, salvation, faith, grace, love, etc.), so it makes sense people would hear from God more clearly and more often.

We have seen three steps that help a person hear from God more clearly and more often. First, if the person is looking to love others, God will want to flow through the person in grace to accomplish this. Since this causes God to send a signal, and we need to be close to the antenna, the second step is to be in your ARE. Finally, our ability to interpret the signal will depend on whether we understand what God means by the words He is using, therefore we need to have non-contradictory definitions.

This book is meant to be read at least twice. Readers have found that on the second reading, it can feel like the book has changed. For example, there is a lot of information in the first five chapters that may have gone over your head. Now when you read those chapters, you may realize you understand the concepts better and will see new information.

The Next Book

The next book is "Modeling God's Wills." That book picks up where this one left off and it isn't that hard to see why. People who believe in the Physical as a driver are trying to accumulate physical resources in order to be happy. They are in pride, thinking they know how to be happy and only lack the means to achieve it.

People who believe that the Spiritual is a driver realize they will only be happy being completely in their ARE and they know they don't know it completely. However, they know three things:

1. God knows them better than anyone else.
2. God loves them more than anyone else.
3. God has all the information.

People with a Spiritual emphasis focus on creating value through justice by handling injustice well and loving with forgiveness. They then give

God access to that value via prayer and God leads them through grace into circumstances that result in happiness.

God's Will is the most complicated doctrine and is the number one reason people become frustrated with their existence. I hope you can see that it would be impossible for anyone to explain God's Will if they couldn't first present a non-contradictory definition of God's Nature.

Understanding God's Will is only possible once a person has understood the information in this book.

This book began with an analogy involving a wall. The goal of this book was to help you continue to move away from the wall when you hit the uncomfortable point where you don't see tiles and you don't see any images. Ideally, some images ought to start becoming clearer. The next book will take you just short of being able to see the entire picture.

Endnotes

1. *Semmelwies, His Life and His Doctrine*, W.J. Sinclair, Manchester University Press, 1909, p. 48-51

2. *Strong's Exhaustive Concordance of the Bible with Hebrew Chaldee and Greek Dictionaries*, James Strong, S.T.D, LL.D., BBH Distributors, A Division of Baker Book House, *A Concise Dictionary of the Words in the Hebrew Bible; With Their Renderings in the Authorized English Version*, "qadash," Hebrew 6942, p. 1021.

3. *History of Science: Antiquity to 1700*, Professor Lawrence M. Principe, The Teaching Company, Lecture Thirty-One: Mechanical Philosophy and Revived Atomism.

4. *Strong's Exhaustive Concordance of the Bible with Hebrew Chaldee and Greek Dictionaries*, James Strong, S.T.D, LL.D., BBH Distributors, A Division of Baker Book House, *A Concise Dictionary of the Words in the Hebrew Bible; With Their Renderings in the Authorized English Version*, "chen," Hebrew 2580, p. 41

5. *Strong's Exhaustive Concordance of the Bible with Hebrew Chaldee and Greek Dictionaries*, James Strong, S.T.D, LL.D., BBH Distributors, A Division of Baker Book House, *A Concise Dictionary of the Words in the Hebrew Bible; With Their Renderings in the Authorized English Version*, "chawan," Hebrew 2603, p. 102

6. *Strong's Exhaustive Concordance of the Bible with Hebrew Chaldee and Greek Dictionaries*, James Strong, S.T.D, LL.D., BBH Distributors, A Division of Baker Book House, *A Concise Dictionary of the Words*

in the Hebrew Bible; With Their Renderings in the Authorized English Version, "charis," Greek 5485, p. 77

7. *Strong's Exhaustive Concordance of the Bible with Hebrew Chaldee and Greek Dictionaries*, James Strong, S.T.D, LL.D., BBH Distributors, A Division of Baker Book House, *A Concise Dictionary of the Words in the Hebrew Bible; With Their Renderings in the Authorized English Version*, "zabach," Hebrew 2076 and 2077 taken together, p. 34

8. Mere Christianity, "Book Two: What Christians Believe," "Chapter 5: The Practical Conclusion," p. 63, HarperSanFrancisco, Harper Collins Edition 2001, Copyright © 1952, C.S. Lewis Pte. Ltd., Copyright renewed © 1980, C.S. Lewis Pte. Ltd.